GitHub Actions
Cookbook

A practical guide to automating repetitive tasks and
streamlining your development process

Michael Kaufmann

GitHub Actions
Cookbook

Group Product Manager: Preet Ahuja
Publishing Product Manager: Prachi Rana
Book Project Manager: Srinidhi Ram
Senior Editor: Sayali Pingale
Technical Editor: Rajat Sharma
Copy Editor: Safis Editing
Indexer: Tejal Soni
Production Designer: Joshua Misquitta
Developer Relations Marketing Executive: Rohan Dobhal

First published: April 2024
Production reference: 1050424

Published by Packt Publishing Ltd.
Grosvenor House
11 St Paul's Square
Birmingham
B3 1RB, UK

ISBN 978-1-83546-894-4
www.packtpub.com

To my family, who had to spend many weekends and nights without me. To my colleagues from Xebia and the DevOps community for giving me feedback, challenging my ideas, and giving me the opportunity to learn.

– Michael Kaufmann

Contributors

About the author

Michael Kaufmann believes that developers and engineers can be happy and productive at work. He loves DevOps, GitHub, Azure, and modern work.

Microsoft has awarded him with the title **Microsoft Regional Director (RD)** and **Microsoft Most Valuable Professional (MVP)** – the latter in the category of DevOps and GitHub.

Michael is also the founder and managing director of Xebia Microsoft Services, Germany – a consulting company that helps its customers become digital leaders by supporting them in their cloud, DevOps, and digital transformation.

Michael shares his knowledge in books, training, and as a frequent speaker at international conferences.

About the reviewers

Mickey Gousset is a staff DevOps architect at GitHub. He is passionate about DevOps and helping developers achieve their goals. Mickey speaks on DevOps and cloud topics at various user groups, code camps, and conferences around the world.

Jamie O'Meara is a passionate technical and business leader with a strong emphasis on software development and data. He is currently a principal solution engineer at GitHub helping customers establish their GitHub solutions and adopt them at scale. GitHub Copilot has piqued his interest in learning more about the use of generative AI. He has been a speaker at various international conferences (VMworld, SpringOne, Cloud Foundry Summit, and more). He is also very interested in the VC community and an avid follower of the start-up ecosystem.

Table of Contents

3

Building GitHub Actions   63

4

The Workflow Runtime   95

5

Automate Tasks in GitHub with GitHub Actions   125

6

Build and Validate Your Code   155

7

Release Your Software with GitHub Actions   197

Preface

GitHub is more than just a platform for hosting and sharing code. With millions of developers from all over the world collaborating on projects of every type and size, it has become the beating heart of the open source community. With GitHub Actions, GitHub now has its own workflow platform that allows engineers and developers to automate all kinds of repetitive engineering tasks – from **Continuous Integration** (**CI**) and **Continuous Deployment** (**CD**) to IssueOps, automatic issue triaging, and ChatOps.

This book will show you how to make the most of GitHub Actions in your day-to-day life. It is a practical book – so we will have you *do* as much as possible and explain the theory alongside the individual recipes.

Who this book is for

If you are looking for a practical approach to learning GitHub Actions, this book is for you, whether you are a software developer or a DevOps engineer. If you have already played around with Actions on your own but want to learn more; you have experience with other CI/CD tools, such as Jenkins or Azure Pipelines; or you are completely new to the topic – it doesn't matter, you'll find this book helpful.

In order to learn from this book, you should have a basic understanding of at least one programming or script language, Git as a version control system, and infrastructure topics such as Docker, the Linux and Windows filesystems, and authentication.

What this book covers

Chapter 1, *GitHub Actions Workflows*, will introduce you to GitHub Actions workflows and what you can do with them. You will learn about **YAML** basics, events that trigger workflows, and expressions, as well as how to use GitHub Actions from the marketplace to automate all kinds of tasks.

Chapter 2, *Authoring and Debugging Workflows*, will teach you best practices for authoring workflows: how to use **Visual Studio Code** and **GitHub Codespaces** and various add-ins to efficiently create, edit, and run workflows, check them for errors with powerful linters, develop them in branches, and run them locally. You will also learn how you can debug them and turn on advanced logging.

Chapter 3, *Building GitHub Actions*, explains the different types of GitHub actions, and you will learn how to use input and output. You will write your own Docker container action, a TypeScript action, and a composite action.

Chapter 4, The Workflow Runtime, is about the different runtime options for your workflows. You will learn how to use different GitHub-hosted runners and how to set up and scale ephemeral, self-hosted runners in Docker containers and Kubernetes with **GitHub Actions Controller** (**GHAC**).

Chapter 5, Automate Tasks in GitHub with GitHub Actions, will show you how to use **Issue-Ops** to automate common tasks within GitHub. You will learn how to authenticate with GitHub Apps, use GITHUB_TOKEN and workflow permissions, use the **GitHub CLI** to automate tasks, use environments for approvals and checks, and use reusable workflows and composite actions to share logic across workflows and repositories.

Chapter 6, Build and Validate Your Code, is about CI. You will learn how to build and test different versions of your code with the same workflow, find security vulnerabilities in your code with CodeQL, attach a **Software Bill of Materials** (**SBOM**) to your release, automate the versioning of your software, and use caching to speed up your workflows.

Chapter 7, Release Your Software with GitHub Actions, covers continuous delivery and continuous deployment. You will learn how to securely deploy to the cloud using **OpenID Connect** (**OIDC**) and how to deploy containers to Kubernetes – whether it is Microsoft **Azure Kubernetes Service** (**AKS**), **Google Kubernetes Engine** (**GKE**), or **Elastic Container Services** (**ECS**). You will also learn how to use Dependabot together with GitHub Actions to completely automate the update of your dependencies.

To get the most out of this book

You will need the following:

Software/hardware covered in the book	OS requirements
GitHub	All operating systems are compatible. You will need an account on https://github.com.
Visual Studio Code	All operating systems are compatible. If you want, you can use GitHub Codespaces for all recipes. In this case, you don't have to install anything locally. If you want to work locally, then you'll need Visual Studio Code (https://code.visualstudio.com/download) and the subsequent tools.
Git	Only required if you work locally. All operating systems are compatible. You should have an up-to-date version of Git installed (at least version 2.23).
GitHub CLI	Only required if you work locally. Install the GitHub CLI (https://cli.github.com/).
Node.js	Only required if you work locally. You'll need an up-to-date version of Node.js (I'm using 21 at the time of writing). All operating systems are compatible. Download the latest version here: https://nodejs.org/en/download/current.

Docker	Only required if you work locally. Get Docker for all operating systems here: `https://docs.docker.com/get-docker/`.
Azure and the Azure CLI	For some chapters, you'll need an Azure account and the Azure CLI. A free test version will be sufficient (`https://azure.microsoft.com/en-us/free`). If you want to work locally, you'll also need the Azure CLI.

All recipes can be done with a free GitHub account in public repositories. You can use GitHub Codespaces to do anything in a virtual environment. This will use up your 120 free hours per month (180 with GitHub Pro). Be aware of that. Once the free hours are used, you'll have to pay per minute for Codespaces.

If you are using the digital version of this book, we advise you to type the code yourself or access the code via the GitHub repository (link available in the next section). Doing so will help you avoid any potential errors related to the copying and pasting of code.

Download the example code files

You can download the example code files for this book from GitHub at `https://github.com/PacktPublishing/GitHub-Actions-Cookbook`. If there's an update to the code, it will be updated in the GitHub repository.

We also have other code bundles from our rich catalog of books and videos available at `https://github.com/PacktPublishing/`. Check them out!

Conventions used

There are a number of text conventions used throughout this book.

`Code in text`: Indicates code words in text, database table names, folder names, filenames, file extensions, pathnames, dummy URLs, user input, and Twitter handles. Here is an example: "We can do this, for example, using the sed `'s/./&/g'` command."

A block of code is set as follows:

```
jobs:
  build-and-push-image:
    runs-on: ubuntu-latest

    permissions:
      packages: write
```

When we wish to draw your attention to a particular part of a code block, the relevant lines or items are set in bold:

```
"dependencies": {
  "@wulfland/package-recipe": "^2.0.5",
```

```
    "express": "^4.18.2"
}
```

Any command-line input or output is written as follows:

```
$ npm start

> release-recipe@1.0.0 start
> node src/index.js

Server running at http://localhost:3000
```

Bold: Indicates a new term, an important word, or words that you see onscreen. For example, words in menus or dialog boxes appear in the text like this. Here is an example: "In your repository, navigate to **Settings | Secrets and Variables | Actions**."

> Tips or important notes
> Appear like this.

Sections

In this book, you will find several headings that appear frequently (*Getting ready*, *How to do it...*, *How it works...*, and *There's more...*).

To give clear instructions on how to complete a recipe, use these sections as follows:

Getting ready

This section tells you what to expect in the recipe and describes how to set up any software or any preliminary settings required for the recipe.

How to do it...

This section contains the steps required to follow the recipe.

How it works...

This section usually consists of a detailed explanation of what happened in the previous section.

There's more...

This section consists of additional information about the recipe in order to make you more knowledgeable about the recipe.

Get in touch

Feedback from our readers is always welcome.

General feedback: If you have questions about any aspect of this book, mention the book title in the subject of your message and email us at customercare@packtpub.com.

Errata: Although we have taken every care to ensure the accuracy of our content, mistakes do happen. If you have found a mistake in this book, we would be grateful if you would report this to us. Please visit www.packtpub.com/support/errata, selecting your book, clicking on the Errata Submission Form link, and entering the details.

Piracy: If you come across any illegal copies of our works in any form on the Internet, we would be grateful if you would provide us with the location address or website name. Please contact us at copyright@packt.com with a link to the material.

If you are interested in becoming an author: If there is a topic that you have expertise in and you are interested in either writing or contributing to a book, please visit authors.packtpub.com.

Share your thoughts

Once you've read *GitHub Actions Cookbook*, we'd love to hear your thoughts! Scan the QR code below to go straight to the Amazon review page for this book and share your feedback.

https://packt.link/r/1835468942

Your review is important to us and the tech community and will help us make sure we're delivering excellent quality content.

Download a free PDF copy of this book

Thanks for purchasing this book!

Do you like to read on the go but are unable to carry your print books everywhere?

Is your eBook purchase not compatible with the device of your choice?

Don't worry, now with every Packt book you get a DRM-free PDF version of that book at no cost.

Read anywhere, any place, on any device. Search, copy, and paste code from your favorite technical books directly into your application.

The perks don't stop there, you can get exclusive access to discounts, newsletters, and great free content in your inbox daily

Follow these simple steps to get the benefits:

1. Scan the QR code or visit the link below

https://packt.link/free-ebook/9781835468944

2. Submit your proof of purchase

3. That's it! We'll send your free PDF and other benefits to your email directly

1

GitHub Actions Workflows

GitHub is more than just a platform for hosting and sharing code. With millions of developers from all over the world collaborating on projects of every type and size, it has become the beating heart of the open source community. Since its foundation in 2008, GitHub has grown to host over 200 million repositories and 100 million users, with a staggering 3.5 billion contributions made in the last year alone. With GitHub Actions, engineers and developers can now automate all kinds of workflows and repetitive engineering tasks – from **Continuous Integration** (**CI**) and **Continuous Deployment** (**CD**) to IssueOps, automatic issue triaging, and ChatOps. GitHub Actions is much more than just a CI/CD tool – it's a comprehensive automation platform that can help streamline your entire development workflow.

This book will show you how to make the most of GitHub Actions in your day-to-day life. It is a practical book – so you will *do* as much as possible, and I will explain the theory alongside the individual recipes.

In this chapter, you will learn the basics of workflows in GitHub: workflow files, the workflow and YAML syntax, events that trigger workflows, expressions, secrets, and environments, and you will write your first workflows.

We're going to cover the following main topics in this chapter:

- The GitHub ecosystem
- Hosting and pricing for GitHub
- Pricing for GitHub Actions
- GitHub Marketplace
- Using the workflow editor for writing workflows
- Using secrets and variables
- Creating and using environments

Technical requirements

For this chapter, you will need a free GitHub account and a browser. Just sign up under `https://github.com/signup` if you do not have an account yet.

You will find all the recipes and example code in the repository at `https://github.com/wulfland/GitHubActionsCookbook`.

The GitHub ecosystem

GitHub is built around the decentralized `git` **version control system** (**VCS**), which has played a significant role in transforming the way in which software is developed. But GitHub is more than just hosting of `git` repositories – it has evolved into a holistic DevOps platform with capabilities in the following areas:

- Collaborative coding
- Planning and tracking
- Workflows and CI/CD
- Developer productivity
- Client applications
- Security

From the very beginning, GitHub has prioritized a developer-centric approach, resulting in a platform that places utmost importance on webhooks and APIs. Developers can leverage either the REST or the GraphQL API to manipulate all aspects of the GitHub platform. In addition to that, developers can use GitHub as an **identity provider** (**IdP**) to access their applications. This approach facilitates seamless integration with other tools and platforms, making GitHub what it is today: the place where the world builds software.

To understand the power of GitHub Actions, one must take into account that you can use it to automate all kinds of tasks in the entire ecosystem – not just code. This includes the following:

- **Planning and tracking**: GitHub offers issues and milestones, GitHub Discussions, and GitHub Projects for planning and tracking. It also integrates seamlessly with other popular planning and tracking solutions such as Jira, Trello, or Azure Boards.

- **Client applications**: GitHub provides Visual Studio Code as a code editor that can be accessed directly in the browser (`https://github.dev`), mobile applications for both iOS and Android platforms, to collaborate from anywhere, a cross-platform desktop application, and has an extensible CLI available.

  It also integrates with all the common IDEs such as Visual Studio, Visual Studio Code, and Eclipse, and with popular chat platforms such as Slack and Teams.

- **Security**: GitHub Advanced Security provides software supply-chain security with Dependabot, Secret Scanning, and code scanning with CodeQL. It also supports integrations with tools such as Snyk, Veracode, or Checkmarx, and it can be integrated into Microsoft's Defender for DevOps.

- **Developer productivity**: GitHub offers a virtual containerized development environment – GitHub Codespaces – and GitHub Copilot, an AI-powered assistant that can help you write and understand code. GitHub also offers code search, a command palette, and other features that can further enhance developer productivity.

- **Workflows and CI/CD**: Beyond GitHub Actions, GitHub supports most CI/CD tools in the market. Furthermore, GitHub provides secure integration with all the major cloud providers for CI/CD workflows using **Open ID Connect (OIDC)**. GitHub Packages provides a package registry that supports a wide range of package formats and native npm support – but all the other major package registries also integrate with GitHub.

GitHub Actions can be used to automate tasks and build solutions across the entire GitHub ecosystem (see *Figure 1.1*):

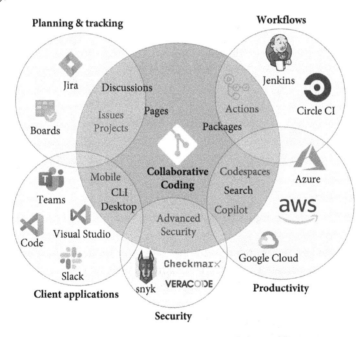

Figure 1.1 – The GitHub ecosystem and its integrations

In this book, I will provide practical recipes for workflows across all the major areas so that you will be able to automate all kinds of real-world development tasks.

Hosting and pricing for GitHub

All the examples in this book are done on `https://github.com` – the **Software-as-a-Service (SaaS)** offering from GitHub. Signing up for GitHub is free and provides users with unlimited private and public repositories. Nearly all features on GitHub are available free for open source projects (public repositories), but they may require a paid license for private repositories. In public repos, you have unlimited minutes for actions. That's why it is important to do all the recipes in public repos – if not, you will burn rapidly through your 2,000 minutes per month.

GitHub's pricing model is based on a monthly per-user billing system and consists of three tiers: **Free**, **Team**, and **Enterprise** (see *Figure 1.2*):

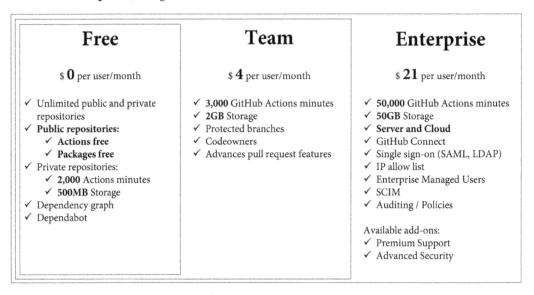

Figure 1.2 – GitHub pricing tiers

As mentioned earlier, public repos are entirely free – including GitHub Actions, Packages, and security features such as Dependabot and Secret Scanning. Private repos are also free, but only with limited functionality for collaboration. It does not include protected branches, Codeowners, and some advanced pull request features. For private repos, you have 2,000 free minutes in the free tier. To unlock the collaboration features, you'll need to acquire a Team license for $4 per user per month. The Team plan then also includes 3,000 minutes for GitHub Actions.

The GitHub Enterprise plan brings all the Enterprise features – such as **single sign-on (SSO)** with **Security Assertion Markup Language (SAML)** and **System for Cross-domain Identity Management (SCIM)**, Enterprise Managed Users, and the IP allow list. It also comes equipped with 50,000 minutes for GitHub Actions – but it also costs $21 per user per month.

In addition to the SaaS offering, GitHub also provides an appliance for self-hosting – **GitHub Enterprise Server (GHES)**. It is available for AWS, GCP, Azure, or on-premises on Hyper-V, OpenStack KVM, or VMware ESXi. GHES is only available with the Enterprise plan. You can also combine GHES with **GitHub Enterprise Cloud (GHEC)** and share the same license for both hosting options.

> **GHES and GitHub Actions**
>
> Keep in mind that you cannot use GitHub-hosted runners for your workflow if you run GHES. You will have to provide your own runners for your workflows and ensure that they are secure and clean up their workflow artifacts. Typically, this is done on Kubernetes with **Actions Runner Controller** (**ARC** – `https://github.com/actions/actions-runner-controller`). You will learn more about this in *Chapter 4, The Workflow Runtime*.

Pricing for GitHub Actions

Running your workflows on self-hosted runners is completely free as you bring your own compute. Running workflows in public repositories is also free – even on the powerful runners provided by GitHub. GitHub-hosted runners are available on Linux, Windows, and macOS and in different sizes. If you want to leverage these runners in private repositories, you'll be charged per minute. The different runners use different **minute multipliers** (see *Table 1.1*). Running a workflow on Linux will reduce 1 of your free minutes per minute – and you will be charged $0.008 if you exceed your free minutes. Windows will burn twice as fast through your free minutes and costs $0.08 per minute after that. And macOS will burn 10 times faster through your minutes and charges $0.016 per minute when you have reached the limit of your included minutes:

Operating system	Minute multiplier	Price per minute
Linux	1	$0.008
Windows	2	$0.080
macOS	10	$0.016

Table 1.1 – Pricing per minute for GitHub-hosted runners

That's the reason why I use Linux for most of the examples in this book and why I always encourage my customers to run as much workload on Linux as possible.

If you use GHEC or the Team plan and you need machines with more power, then you can leverage larger GitHub-hosted runners. They are charged by minute (see *Table 1.2*) and have additional features such as static IP ranges:

vCPUs	Linux	Windows	macOS
2	$0.008	$0.016	
3			$0.08
4	$0.016		
8	$0.032	$0.064	
12			$0.32
16	$0.064	$0.128	
32	$0.128	$0.256	
64	$0.256	$0.512	

Table 1.2 – Per-minute rate for larger runners

> **Private networking**
>
> In addition to static IP ranges, you can also use Azure private networking to connect GitHub-hosted runners directly to your resources. At the time of writing, this feature is still in beta and might change. See the following link for more information: `https://docs.github.com/en/enterprise-cloud@latest/admin/configuration/configuring-private-networking-for-hosted-compute-products/about-networking-for-hosted-compute-products`.

GitHub Actions also consumes storage – for example, for logs, workflow artifacts, or caching. If you exceed your included storage, you will be billed $0.008 per GB per day.

Keep in mind that pricing may change, and refer to the GitHub documentation for up-to-date information (`https://docs.github.com/en/billing/managing-billing-for-github-actions/about-billing-for-github-actions`).

And to learn GitHub Actions and experiment with workflows – just do everything in public repositories and you will not have to pay, either for compute or for storage.

GitHub Marketplace

GitHub offers a community-driven marketplace (`https://github.com/marketplace`) that contains currently more than 20,000 GitHub Actions that you can reuse as building blocks in your workflows (see *Figure 1.3*):

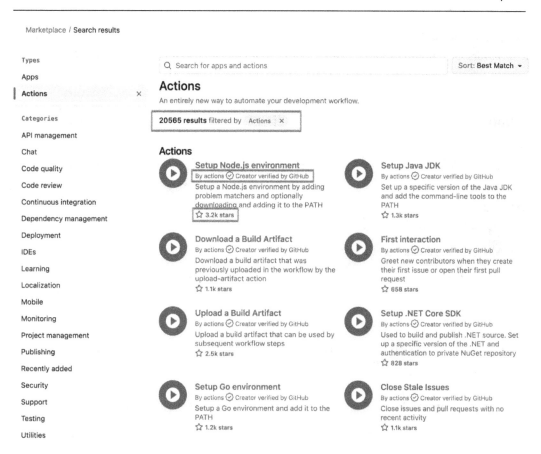

Figure 1.3 – GitHub Marketplace contains more than 20,000 reusable actions

If an action is by the author *actions*, that means it is a native action by GitHub. You can see the number of people who have starred an action in the overview. This will give you a good indication of the popularity of the action. And you will see the blue badge that indicates that the author of an action was verified by GitHub.

You can filter the marketplace by multiple categories, you can search by terms, and you can change the sort order of results to **Most installed/starred**, **Best Match**, or **Recently added** (see *Figure 1.4*):

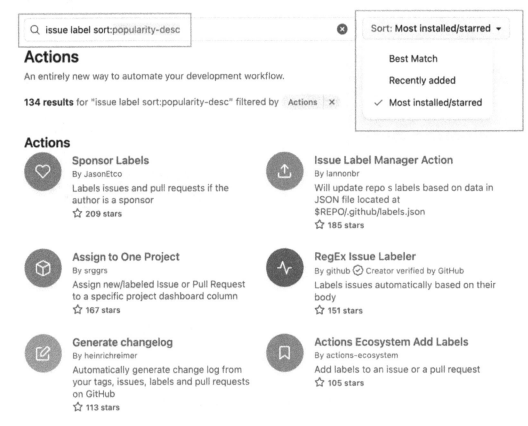

Figure 1.4 – Searching in the marketplace and sorting the results

This way, it is easy to explore the marketplace and find actions that will help you automate tasks in your workflows.

If you click one of the results, it will take you to the details page of the marketplace listing (see *Figure 1.5*):

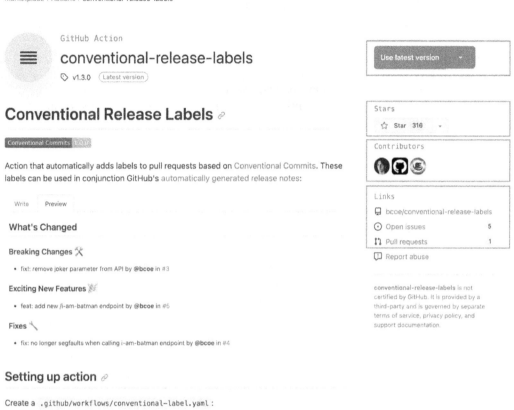

Figure 1.5 – Details of a marketplace listing

You can find released versions, the number of stars, contributors, and – as all published actions are open source – a link to the source repository and the number of open issues and pull requests. This should give you a good idea of how actively the action is used – and it allows you to dig into the code if you wish to do so.

The results of the marketplace are also displayed in the workflow editor, and we will use them in our recipes from it.

Using the workflow editor for writing workflows

GitHub does a good job of guiding people in the workflow designer when writing workflows. That's why it is best to just start and write your first workflow and familiarize yourself with the platform.

Getting ready

Before you can create your first workflow, you first have to create a repository on GitHub. Navigate to `https://github.com/new`, authenticate if you are not authenticated yet, and fill in data as in *Figure 1.6*:

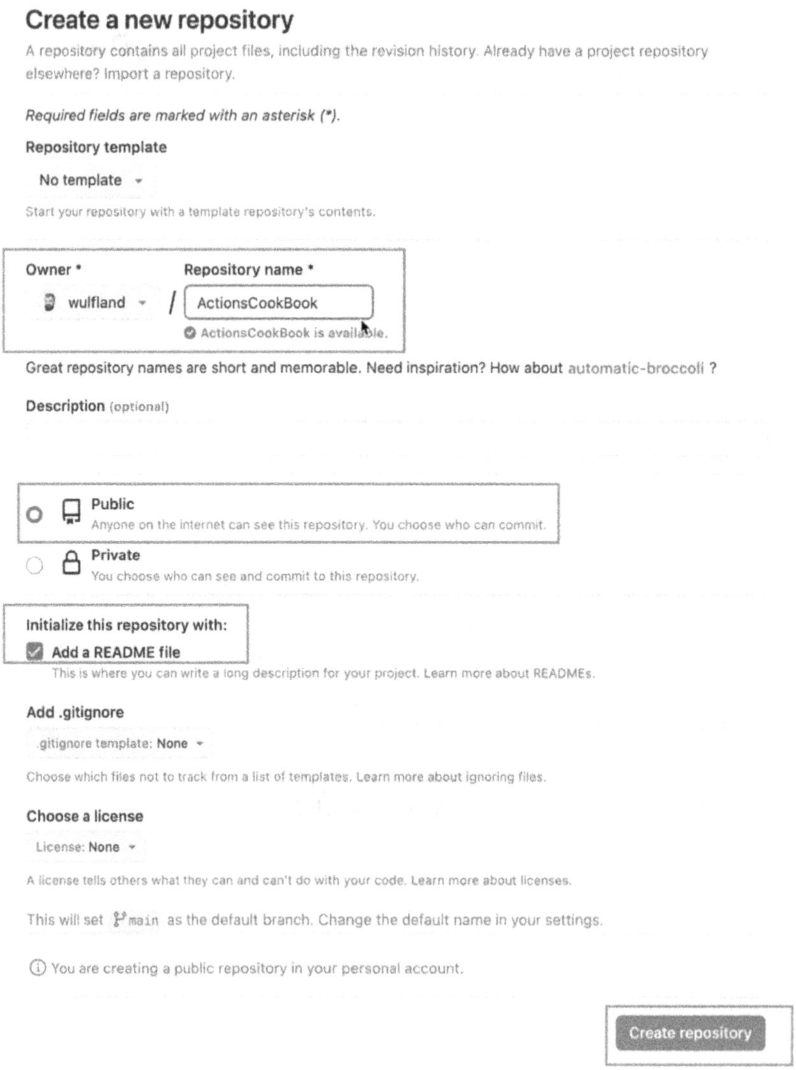

Figure 1.6 – Creating a new repository

Pick your GitHub user as the owner and give the repo a unique name – for example, `ActionsCookBook`. Make it a public repo so that all workflows and storage are free. Initialize the repo with a README file – this way, we have already files in the repo and something in the workflow to work with.

How to do it...

GitHub Action workflows are **YAML** files with a `.yml` or `.yaml` extension that are located in the `.github/workflows` folder in a repository. You could create the file manually, but then the workflow editor would only work after the first commit. Therefore, I recommend creating a new workflow from the menu.

1. In your new repository, navigate to **Actions**. Since your repository is new and you don't have any workflows yet, this will redirect you directly to the **Create new workflow** page (`actions/new`). If your repository contains workflows, you will see the workflows here (as later displayed in *Figure 1.16*), and you would have to click the **New workflow** button to get to that page.

    On this page, you will find a lot of template workflows you could use as a starting point. There are starter workflows for *deployments* to most clouds, *CI* for most languages, *security* scanning of your code, *automation* in general, and templates to deploy content to GitHub Pages. You can filter the starter workflows by these categories. These workflows give you a good starting point for most of your workflows.

    In this recipe, we will focus on familiarizing ourselves with the editor, and we will create a workflow from scratch by clicking **set up a workflow yourself** (see *Figure 1.7*):

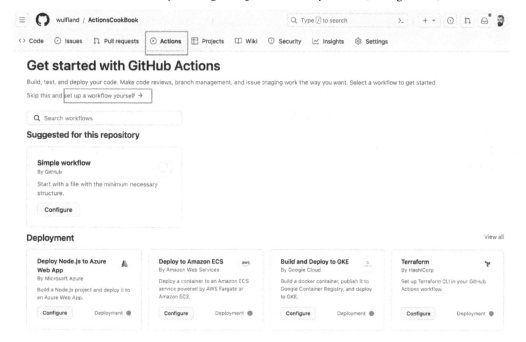

Figure 1.7 – Creating a new workflow in GitHub

2. GitHub will create a new `main.yml` file in `.github/workflows` on the default branch and display it in the web editor. On the right side of the editor, you have the documentation, and you can search for `actions` in GitHub Marketplace. In the editor, you can use *Ctrl + Space* (or *Option + Space* – depending on your keyboard settings) to trigger autocomplete. The editor will capture the *Tab* key and by default use it for a two-space indentation. To navigate to other controls on the page using the *Tab* key, you first have to exit it using *Esc* or using *Ctrl + Shift + M*.

Modify the filename to `MyFirstWorkflow.yml` and familiarize yourself with the editor (see *Figure 1.8*):

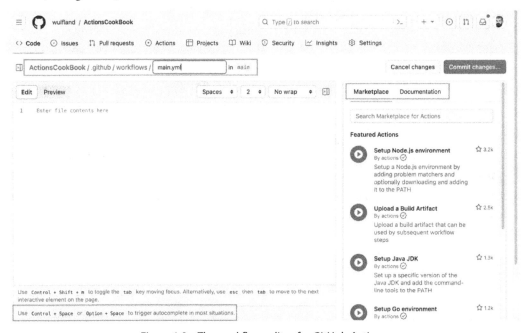

Figure 1.8 – The workflow editor for GitHub Actions

3. In the editor, click *Ctrl + Space* (or *Option + Space*) to see a list of root elements valid in a workflow file (see *Figure 1.9*):

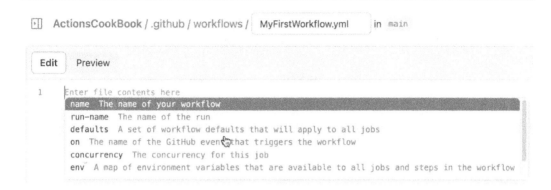

Figure 1.9 – The editor shows you all valid options at a certain level in the workflow file

Typically, workflows are started with the name property, which sets the display name of the workflow in the UI. It's a good practice to add a comment to the top of the file summarizing the intent of the workflow.

4. Add a comment to the top of the file and set the name property using autocomplete. Note that the editor has error checking and indicates that you are still missing the required root key, on (see *Figure 1.10*):

Figure 1.10 – Error checking in the code editor

5. Next, we are going to configure events that should trigger the workflow. Note that a workflow can have multiple triggers. Depending on where you are in the designer, autocomplete will give you different results. If you are on the same line as on:, you will get a result in the JSON syntax (see the *YAML collection types* section); that is, on: [push].

If you add a comma after the first element and click *Control + Space* again, then you can pick additional elements from autocomplete (see *Figure 1.11*):

Figure 1.11 – Autocomplete works also inside squared brackets

Each trigger is a map and can contain additional arguments. If you put your cursor on the line below on: and add a two-space indentation, autocomplete will give you the results in the full YAML syntax. It will also give you properties that you can use to configure each trigger (see *Figure 1.12*):

Figure 1.12 – Autocomplete also helps with options for triggers

Note that most arguments – for example, branches or paths – are sequences and need a dash for each entry if you are not using the JSON syntax.

We want our test workflow to run on every push to the `main` branch. We also want to be able to trigger it manually (see the *Events that trigger workflows* section). Your workflow code for triggers should look like this:

```
on:
  push:
    branches:
      - main
  workflow_dispatch:
```

Wildcards

The * character can be used as a wildcard in paths and ** as a recursive wildcard. * is a special character in YAML, so you need to use quotation marks in that case:

```
push:
    branches:
      - 'release/**'
    paths:
      - 'doc/**'
```

6. After configuring the triggers for the workflow, the next step is to add another root element: the *jobs*. Jobs are a map in YAML – meaning on the next line with two-space indentation, autocomplete will not work as the editor expects you to set a name. Name your job `first_job` and go to the next line. The name of the job object can only contain alphanumeric values, a dash (-), and an underscore (_). If you want any other characters to be displayed in the workflow, you can use the `name` property:

```
jobs:
  first_job:
    name: My first job
```

7. Every job needs a runner that executes it. Runners are identified by labels. You will learn more about runners in *Chapter 4, The Workflow Runtime*. We want our workflow to be executed on the latest version of the Ubuntu runners provided by GitHub, so we use the `ubuntu-latest` label:

```
runs-on: ubuntu-latest
```

8. A job consists of a sequence of steps that are executed one after the other. The most basic step is the `run:` command, which will execute a command-line command:

```
steps:
    - name: Greet the user
```

```
    run: echo "Hello world"
    shell: bash
```

The name is optional and sets the output of the step in the log. The shell is also optional and will default to bash on non-Windows platforms, with a fallback to sh. On Windows, the default is PowerShell Core (pwsh), with a fallback to cmd. But you could configure any shell you want with the {0} placeholder for the input of the step (that is, shell: perl {0}).

To add variable output, we can use **expressions** that are written between ${{ and }}. In the expression, you can use values from context objects such as the GitHub context. Note that autocomplete also works for these context objects (see *Figure 1.13*):

```
jobs:
  first_job:
    name: My first job
    runs-on: ubuntu-latest
    steps:
      - run: echo "Hello world 🌏 from ${{ github.|
                                        event
                                        workflow
                                        actor
                                        repository
                                        event_name
                                        sha
```

Figure 1.13 – Autocomplete also works for context objects

Pick the actor from the list of values:

```
    - run: echo "Hello world 🌏 from ${{ github.actor }}."
```

You will learn more about expressions and context syntax throughout the book. But you can refer to the documentation for expressions (https://docs.github.com/en/actions/learn-github-actions/expressions) and context (https://docs.github.com/en/actions/learn-github-actions/contexts) at any time.

9. YAML allows you to write multiline scripts without the need to wrestle with quotations and newlines. Just add the pipe operator (|) after run: and write your script in the next line with a four-space indentation. YAML will treat this as one block until the next element – even with new and blank lines in it:

```
    - run: |
        echo "Hello world 🌏 from ${{ github.actor }}."

        echo "Current branch is '${{ github.ref }}'."
```

10. GitHub Actions workflows will not automatically download the code from your repository. If you want to do something with files in your repository, you have to check out the content first. This is done using a GitHub action – a reusable workflow step that can easily be shared for multiple workflows.

On the right side of the workflow editor is the **marketplace**. You can directly search there for all kinds of actions. Search for `checkout` and locate the action from **actions** (these are built-in actions from GitHub). In the listing, you see the owner of the action, the latest version, and the stars of the repository. The listing contains an **Installation** section that you can copy into your workflow to use the action (see *Figure 1.14*):

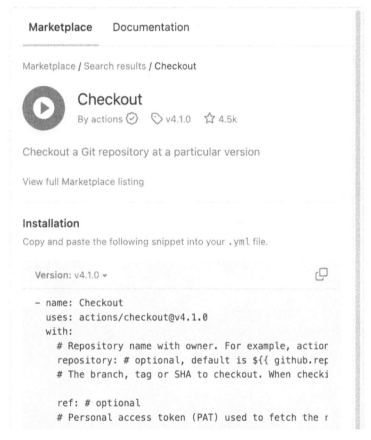

Figure 1.14 – Listing of the marketplace in the workflow editor

Note that many parameters are optional. To check out the repo, you only need the following lines:

```
- name: Checkout
  uses: actions/checkout@v4.1.0
```

> **Using GitHub Actions**
>
> Actions refer to a location on GitHub. The syntax is `{path}@{ref}`. The path points to a physical location on GitHub and can be `{owner}/{repo}` if the actions are in the root of a repository or `{owner}/{repo}/{path}` if the actions are in a subfolder. The reference after `@{ref}` is any `git` reference that points to a commit. It can be a **tag**, **branch**, or an individual commit **SHA**.

11. To display the files in our repository after checking them out, we'll add an extra step:

    ```
    - run: tree
    ```

    This will output the files in the repository in a tree structure.

12. To run the workflow, just commit the workflow file to the `main` branch. Click **Commit changes…**, leave the commit message and branch, and click **Commit changes** in the dialog to finish the operation (see *Figure 1.15*):

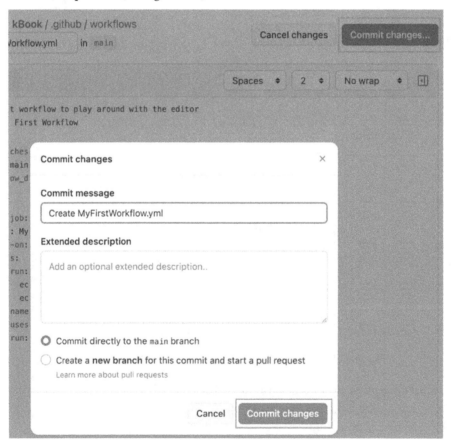

Figure 1.15 – Committing the workflow file

13. As we have set a push trigger for the `main` branch, our commit has automatically triggered the workflow. If you navigate now to **Actions** in your repository, you will be able to see your workflow and the latest workflow run (see *Figure 1.16*):

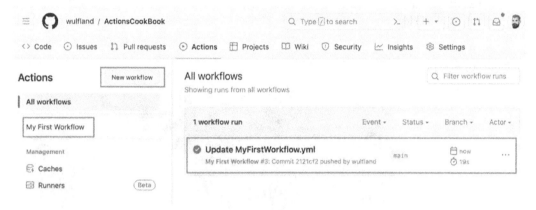

Figure 1.16 – The default view in Actions displays the latest workflow runs of all workflows

Note that the name of the workflow run is the commit message. You can also see the commit that triggered the workflow and the actor that pushed the changes.

14. Click on the workflow run to see more details. The workflow summary page contains jobs on the left and a visual representation on the right (see *Figure 1.17*). It also contains metadata for the trigger, the status, and the duration:

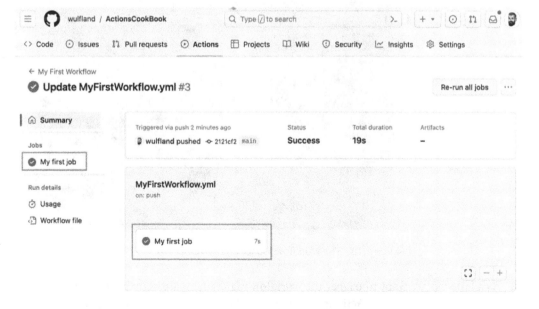

Figure 1.17 – The workflow summary page

15. Click on the job to view more details. In the workflow log, you can inspect the individual steps. Note that each line of the workflow file has a clickable number – that is a URL that you could use to identify each line. The **Set up job** step is a special step that gives you a lot of background information about the workflow runner and workflow permissions (see *Figure 1.18*). Inspect the output of all steps of your workflow:

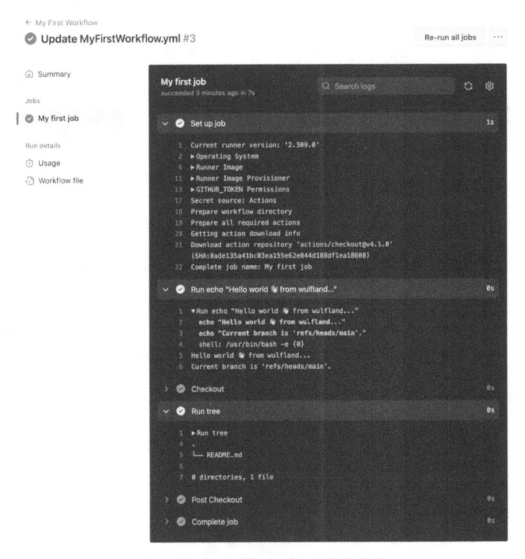

Figure 1.18 – The workflow log for an individual job

16. As a last step, we want to trigger the workflow manually to also see the difference in the workflow run. Go back to **Actions**, select the workflow on the left side (see *Figure 1.19*), and run the workflow:

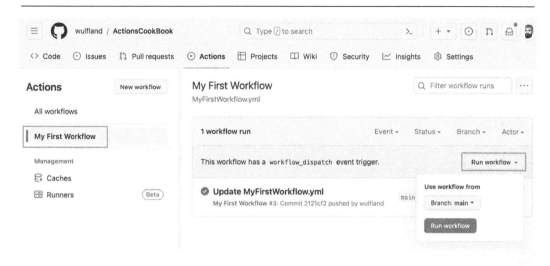

Figure 1.19 – Triggering a workflow manually through the UI

Inspect the new workflow run and its output.

How it works...

Workflow files are YAML files located in the `.github/workflows` folder in a repository.

YAML basics

YAML stands for **YAML Ain't Markup Language** and is a data-serialization language optimized to be directly writable and readable by humans. It is a strict superset of JSON but with syntactically relevant newlines and indentation instead of braces.

You can write comments by prefixing text with a hash (#).

In YAML, you can assign a value to a variable with the following syntax: `key:  value`.

`key` is the name of the variable. Depending on the data type of `value`, the type of the variable will be different. Note that keys and values can contain spaces and do not need quotation marks! Only add them if you use some special characters or you want to force certain values to be a string. You can quote keys and values with single or double quotes. Double quotes use the backslash as the escape pattern (`"Foo \"bar \"  foo"`), while single quotes use an additional single quote for this (`'foo ''bar'' foo'`).

YAML collection types

In YAML, there are two different collection types: nested types called maps and lists – also called sequences. Maps use two spaces of indentation:

```
parent_type:
  key1: value1
```

```
key2: value2
nested_type:
  key1: value1
```

A sequence is an ordered list of items and has a dash before each line:

```
sequence:
  - item1
  - item2
  - item3
```

Since YAML is a superset of JSON, you can also use the JSON syntax to put collections in one line:

```
key: [item1, item2, item3]
key: {key1: value1, key2: value2}
```

Events that trigger workflows

There are three types of triggers for workflows: *webhook triggers*, *scheduled triggers*, and *manual triggers*.

Webhook triggers start the workflow based on an event in GitHub. There are many webhook triggers available. For example, you could run a workflow on an `issues` event, a `repository` event, or a `discussions` event. The `push` trigger in our example is a webhook trigger.

Scheduled triggers can run the workflow at multiple scheduled times. The syntax is the same syntax used for `cron` jobs:

```
on:
  schedule:
    # Runs at every 15th minute
    - cron:  '*/15 * * * *'
    # Runs every hour from 9am to 5pm
    - cron:  '0 9-17 * * *'
    # Runs every Friday at midnight
    - cron:  '0 2 * * FRI'
```

Manual triggers allow you to start the workflow manually. The `workflow_dispatch` trigger will allow you to start the workflow using the web UI or GitHub CLI. You can define input parameters for this trigger using the `inputs` property. The `repository_dispatch` trigger can be used to trigger the workflow using the API. This trigger can also be filtered by certain event types and can accept additional JSON payload that can be accessed in the workflow.

To learn more about triggers, check the documentation at `https://docs.github.com/en/actions/using-workflows/events-that-trigger-workflows`.

Jobs

Every job needs a runner that executes it. Runners are identified by labels. In our recipe, we use the ubuntu-latest label. This means that our job will be executed on the latest Ubuntu image hosted by GitHub. You will learn more about runners in *Chapter 4, The Workflow Runtime*.

Using GitHub Actions

Actions refer to a location on GitHub. The syntax is {path}@{ref}. The path points to a physical location on GitHub and can be {owner}/{repo} if the actions are in the root of a repository or {owner}/{repo}/{path} if the actions are in a subfolder. The reference after @{ref} is any git reference that points to a commit. It can be a tag, branch, or an individual commit SHA:

```
# Reference a version using a tag
- uses: actions/checkout@v4.1.0

# Reference the current head of a branch
- uses: actions/checkout@main

# Reference a specific commit
- uses: actions/checkout@8e5e7e5ab8b370d6c329ec480221332ada57f0ab
```

For local actions in the same repository, you can omit the reference if you check out of the repository.

If the action has defined inputs, you can specify them using the with property:

```
- uses: ActionsInAction/HelloWorld@v1
  with:
    WhoToGreet: Mona
```

Inputs can be optional or required. You can also set environment variables for steps using the env property:

```
- uses: ActionsInAction/HelloWorld@v1
  env:
    GITHUB_TOKEN: ${{ secrets.GITHUB_TOKEN }}
```

There's more...

This is just a very basic workflow that uses an action to check out code and runs some commands on the command line. In the next two recipes, I'll show you how to use secrets, variables, and protected environments for more complex workflows.

Using secrets and variables

You can set variables and secrets in a repository that you can access in workflows. In this recipe, we'll add both and access them in the workflow.

Getting ready

In this recipe, we will use the web UI to set variables and secrets. You can also use the GitHub CLI (`https://cli.github.com/`) for that. If you want to try that, then you have to install it. But it is not necessary for following the recipe.

How to do it...

1. In your repository, navigate to **Settings | Secrets and Variables | Actions**. You can see all existing secrets in the repository, and you can toggle the tabs between **Secrets** (`settings/secrets/actions`) and **Variables** (`settings/variables/actions`; see *Figure 1.20*):

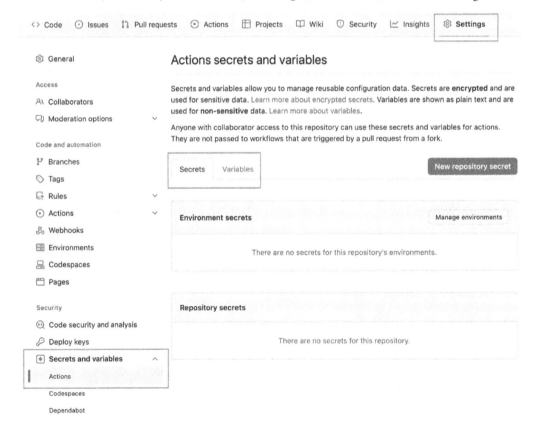

Figure 1.20 – Configuring secrets and variables for a repository

2. Clicking **New repository secret** will open the **New secret** dialog (`settings/secrets/actions/new`; see *Figure 1.21*):

Actions secrets / New secret

Name *

```
YOUR_SECRET_NAME
```

Secret *

```



```

Add secret

Figure 1.21 – Adding a new secret

Add `MY_SECRET` as the secret name and a random word such as `Abracadabra` as the secret, and click **Add secret**. The secret will be masked in the logs! So, don't use a common word that could occur in other outputs of random jobs or steps.

> **Naming conventions for secrets and variables**
>
> Secret names are not case-sensitive, and they can only contain normal characters (`[a-z]` and `[A-Z]`), numbers (`[0-9]`), and an underscore (`_`). They must not start with `GITHUB_` or a number.
>
> The best practice is to name secrets with uppercase words separated by the underscore character.

3. Repeat the process for **New repository variable** (`settings/variables/actions/new`) and create a `WHO_TO_GREET` variable with the value `World`.

4. Open the `.github/workflows/MyFirstWorkflow.yml` file from the previous recipe and click the edit icon (see *Figure 1.22*):

Figure 1.22 – Editing MyFirstWorkflow.yml

Change the word `World` to the `${{ vars.WHO_TO_GREET }}` expression and add a new line using the `${{ secrets.MY_SECRET }}` secret:

```
- run: |
    echo "Hello ${{ vars.WHO_TO_GREET }} 👋 from ${{ github.
actor }}."
    echo "My secret is 🕵 ${{ secrets.MY_SECRET }}."
```

5.   Commit the changes. The workflow will run automatically. Inspect the output in the workflow log. It should look like *Figure 1.23*:

Figure 1.23 – Output of a secret and variable in the log

There's more...

You can create configuration variables for use across multiple workflows by defining them on one of the following levels:

- Organization level
- Repository level
- Environment level

The three levels work like a hierarchy: you can override a variable or secret on a lower level by providing a new value to the same key. *Figure 1.24* illustrates the hierarchy:

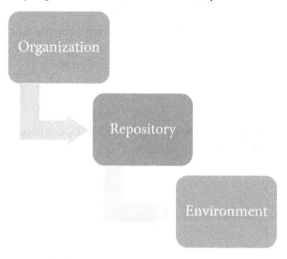

Figure 1.24 – The hierarchy for configuration variables and secrets

Secrets and variables for organizations work the same way as for repositories. You can create a secret or variable under **Settings | Secrets and variables | Actions**. New organization secrets or variables can have an access policy for the following:

- **All repositories**
- **Private repositories**
- **Selected repositories**

When choosing **Selected repositories**, you can grant access to individual repositories.

In addition to setting these values through the UI, it is also possible to use the GitHub CLI.

You can use `gh secret` or `gh variable` to create new entries:

```
$ gh secret set secret-name
$ gh variable set var-name
```

You will be prompted for the secret or variable values, or you can read the value from a file, pipe it to the command, or specify it as the body (`-b` or `--body`):

```
$ gh secret set secret-name < secret.txt
$ gh variable set var-name --body config-value
```

Creating and using environments

Environments are used to describe a general deployment target such as `development`, `test`, `staging`, or `production`. You can protect environments with protection rules, and you can provide configuration variables and secrets for specific environments.

Getting ready

We will first create some environments using the web UI and add some protection rules, secrets, and variables. Then, we add them to our existing workflow.

How to do it...

1. Navigate to **Settings | Environments** and click on **New environment** (see *Figure 1.25*):

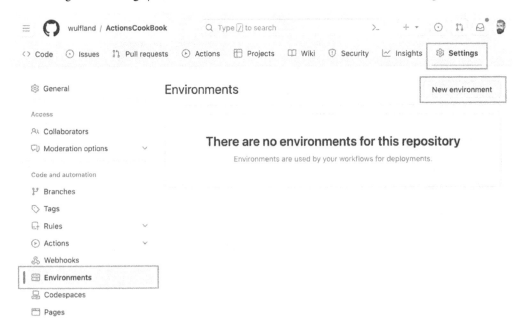

Figure 1.25 – Managing environments in a repository

Enter the name `Production` and click **Configure environment** (see *Figure 1.26*):

Environments / **Add**

Name *

Production

Configure environment

Figure 1.26 – Creating a new environment

2. Add yourself as a required reviewer and click **Save protection rule** (see *Figure 1.27*):

Environments / **Configure Production**

Deployment protection rules

Configure reviewers, timers, and custom rules that must pass before deployments to this environment can proceed.

☑ **Required reviewers**
Specify people or teams that may approve workflow runs when they access this environment.

Add up to 5 more reviewers

Search for people or teams...

🖼 wulfland   ✕

☐ **Wait timer**
Set an amount of time to wait before allowing deployments to proceed.

Enable custom rules with GitHub Apps (Beta)
Learn about existing apps or create your own protection rules so you can deploy with confidence.

☑ Allow administrators to bypass configured protection rules

Figure 1.27 – Configuring deployment protection rules

3. Under **Deployment branches and tags**, choose **Selected branches and tags**, click the plus symbol, and add a name pattern for the `main` branch (see *Figure 1.28*):

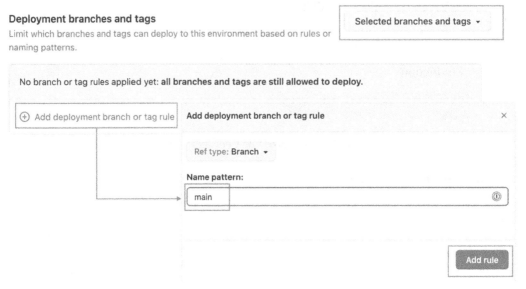

Figure 1.28 – Configuring deployment branches and tags

4. Under **Environment secrets**, click on **Add secret** and add a new `MY_SECRET` secret with the value `Open Sesame` (see *Figure 1.29*). Repeat this with **Add variable** and add a `WHO_TO_GREET` variable with the value `Production users`:

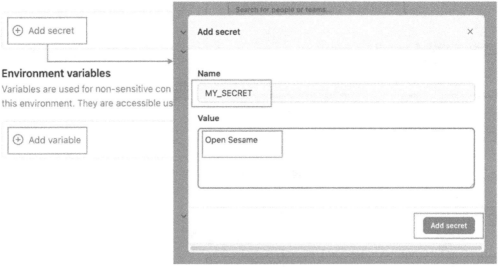

Figure 1.29 – Adding secrets and variables to environments

5. Repeat *step 1* and create two additional environments, `Test` and `Load-Test`. We will use these environments in the next steps to show how to execute jobs in parallel. You don't have to configure deployment branches or required reviewers. Just add a `WHO_TO_GREET` variable with the corresponding value. The result should look like *Figure 1.30*:

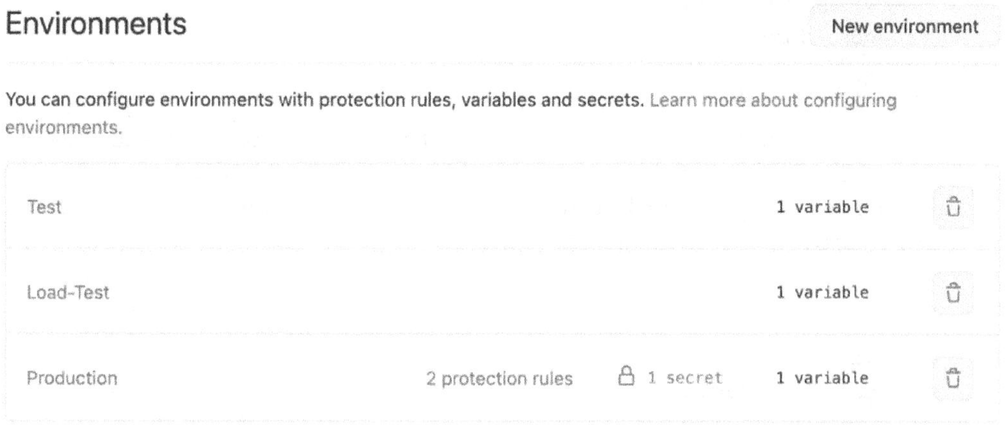

Environments

New environment

You can configure environments with protection rules, variables and secrets. Learn more about configuring environments.

Test			1 variable	🗑
Load-Test			1 variable	🗑
Production	2 protection rules	🔒 1 secret	1 variable	🗑

Figure 1.30 – Multiple environments in the settings of the repository

6. Now, go back to the workflow file and edit it. Add a new job beneath `first_job` called `Test` that runs on the latest Ubuntu image. We associate this job with the `Test` environment. To run this job after `first_job`, we use the `needs` property and set it to the job we depend on:

```
Test:
    runs-on: ubuntu-latest
    environment: Test
    needs: first_job
```

To see how secrets are overwritten by the environment, we have to use a little hack. As GitHub searches for the value of secrets in the output of the log to mask it, we have to modify the actual text. We can do this, for example, using the `sed 's/./& /g'` command. This will add a blank between every character of the secret. With this little hack, the steps of the `Test` job should look like this:

```
steps:
- run: |
    echo "Hello ${{ vars.WHO_TO_GREET }} 👋 from ${{ github.
actor }}."
    sec=$(echo ${{ secrets.MY_SECRET }} | sed 's/./& /g')
    echo "My secret is 🙊 '$sec'."
```

7. Next, add a new `Load-Test` job that is associated with the `Load-Test` environment and also executes after `first_job`:

```
Load-Test:
  runs-on: ubuntu-latest
  environment: Load-Test
  needs: first_job
```

Just copy the steps from `Test`. There is no need to change anything.

8. The last job is a `Production` job. In addition to the name, the `environment` property accepts a URL that later will be displayed in the workflow designer. Set it to any URL you want. To show how after a parallel execution of jobs the workflow can merge again, we will run `Production` after Test and Load-Test:

```
Production:
  runs-on: ubuntu-latest
  environment:
    name: Production
    url: https://writeabout.net
  needs: [Test, Load-Test]
```

Just copy the steps from the previous jobs.

9. Commit your changes to the `main` branch. The workflow will run automatically. Navigate to the new workflow run and inspect the workflow designer, which nicely shows the parallel execution. The workflow will pause before executing `Production` and will wait for approval (see *Figure 1.31*):

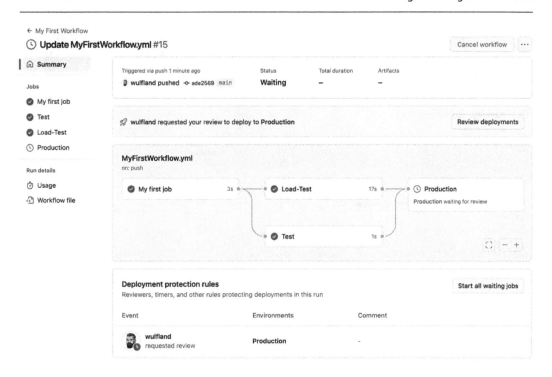

Figure 1.31 – The workflow will stop before an environment with required reviewers and wait for approval

10. Click **Review deployment**, check **Production**, and add an optional comment. Click **Approve and deploy** to start executing the `Production` job (see *Figure 1.32*):

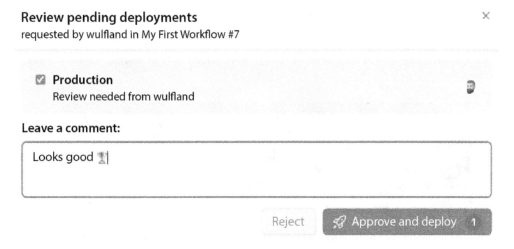

Figure 1.32 – Approving a protected environment

The workflow will execute completely, and the result should look like *Figure 1.33*. Note that the URL is displayed in the **Production** environment. Also, note the history of approvals in the workflow summary:

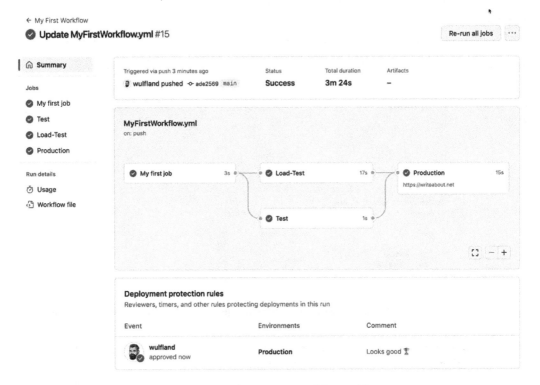

Figure 1.33 – The final summary of the workflow

Open the individual jobs and inspect the output of the step we added (see *Figure 1.34*). The secrets and variables are used from the repository and are only overridden if we set them in an environment:

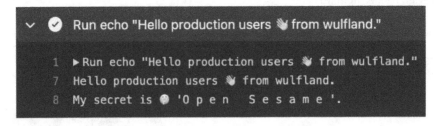

Figure 1.34 – The production secret is only available to the production environment after approval

There's more...

If you are setting secrets or variables for an environment using the GitHub CLI, then you can specify them using the `--env` (`-e`) argument. For organization secrets, you set the visibility (`--visibility` or `-v`) to `all`, `private`, or `selected`. For `selected`, you must specify one or more repos using `--repos` (`-r`):

```
$ gh secret set secret-name --env environment-name
$ gh secret set secret-name --org org -v private
$ gh secret set secret-name --org org -v selected -r repo
```

Environments have more options than we have used in this recipe. You can also configure a wait timer that will pause the workflow for n minutes (with a maximum of 30 days) before executing the deployment job for that particular environment.

There is also a new feature called *custom deployment protection* rules that is still in beta. This feature allows the creation of GitHub apps that can pause your deployment and wait for a specific condition. There are already apps from Datadog, Honeycomb, Sentry, New Relic, and ServiceNow (see https://docs. github.com/en/actions/deployment/protecting-deployments/configuring-custom-deployment-protection-rules#using-existing-custom-deployment-protection-rules). We'll have a closer look at custom deployment rules in *Chapter 7, Release Your Software with GitHub Actions*.

The true power of environment protection rules lies in the **deployment branch or tag rules**. This can restrict code that does not apply to branch protection rules from deploying to certain environments. This can include all kinds of checks – Codeowners approvals, code reviewers, deployments to certain other environments, SonarQube quality gates, and many other automated code checks (see https:// docs.github.com/en/repositories/configuring-branches-and-merges-in-your-repository/managing-protected-branches/about-protected-branches for more information).

2

Authoring and Debugging Workflows

This chapter goes a step further and you will learn best practices for authoring workflows. This includes using Visual Studio Code, running your workflows locally, linting, working in branches, and using advanced logging and monitoring. This will be the foundation for the other chapters, as it gives you plenty of options on how to write your workflows.

This chapter covers the following:

- Using Visual Studio Code for authoring workflows
- Developing workflows in branches
- Linting workflows
- Writing messages to the log
- Enabling debug logging
- Running your workflows locally

Technical requirements

For this chapter, you need **Visual Studio Code** (**VS Code**) installed on your local machine. It is available for Windows (x64, x86, and Arm64), Linux (x64, x86, and Arm64), and Mac (Intel and Apple silicon), and you can install it from the following website if you haven't already done this: `https://code.visualstudio.com/download`.

Additionally, check that you have an up-to-date Git version installed on your machine. You can get instructions on how to get the Git client here: `https://git-scm.com/downloads`.

To run your workflows locally, you will also need **Docker** installed. If you are using macOS, please be sure to follow the steps outlined in **Docker Docs** for how to install Docker Desktop for Mac (`https://docs.docker.com/docker-for-mac/install`). If you are using Windows, please follow steps for installing **Docker Desktop** on Windows (`https://docs.docker.com/docker-for-windows/install`). If you are using Linux, you will need to install **Docker Engine** (`https://docs.docker.com/engine/install`).

Using Visual Studio Code for authoring workflows

Visual Studio Code (**VS Code**) is one of the most popular and widely used code editors in the world. It has gained significant popularity in the developer community due to its flexibility, extensive ecosystem of extensions, and strong community support.

VS Code has a high level of integration with GitHub. It offers features such as Git integration, the synchronization of your settings using your GitHub account, direct access to repositories, and the ability to create, edit, and manage GitHub Action workflows from within the editor using the extension provided by GitHub. This tight integration simplifies the workflow creation process and streamlines collaboration on GitHub action workflows.

In this recipe, we'll install the VS Code extension for GitHub Actions and inspect what you can do with it.

Getting ready...

Before we begin, check that your email address and name are set correctly in Git:

```
$ git config --global user.email
$ git config --global user.name
```

Keep in mind that we work in public repositories. If you want to keep your email address private, use the mail address from `https://github.com/settings/emails` (see *Figure 2.1*):

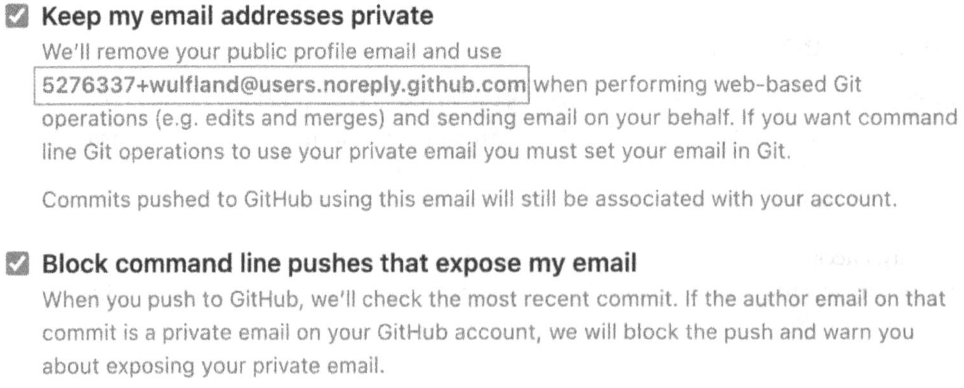

☑ **Keep my email addresses private**
We'll remove your public profile email and use
5276337+wulfland@users.noreply.github.com when performing web-based Git
operations (e.g. edits and merges) and sending email on your behalf. If you want command
line Git operations to use your private email you must set your email in Git.

Commits pushed to GitHub using this email will still be associated with your account.

☑ **Block command line pushes that expose my email**
When you push to GitHub, we'll check the most recent commit. If the author email on that
commit is a private email on your GitHub account, we will block the push and warn you
about exposing your private email.

Figure 2.1 – Keep your email address private in public repositories

This email address consists of your GitHub user ID and name in the `users.noreply.github.com` domain:

```
$ git config --global user.email 5276337+wulfland@users.noreply.
github.com
```

GitHub will automatically associate your commit with your account without exposing your email address.

Then, clone your repository from *Chapter 1* locally. You can find the corresponding URLs in the **Code** section of your repository under **Code | Local** (see *Figure 2.2*):

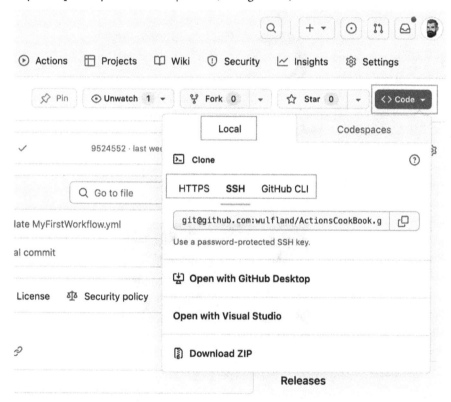

Figure 2.2 – Cloning your repository locally

I used SSH to authenticate, as I can manage my SSH keys in 1Password—but you can also use HTTPS and a PAT token. You can find more information on cloning repositories locally here: `https://docs.github.com/en/repositories/creating-and-managing-repositories/cloning-a-repository`.

How to do it...

1. Open VS Code. Open the extensions window by typing *[Shift]+[Command]+[X]* or *[Ctrl]+[Shift]+[X]* or by just clicking the **Extension** icon in the left bar (see *Figure 2.3*). Search for `github actions` and install the action from GitHub with the verified badge (`https://marketplace.visualstudio.com/items?itemName=GitHub.vscode-github-actions`). Restart VS Code if necessary, and sign in with your GitHub account:

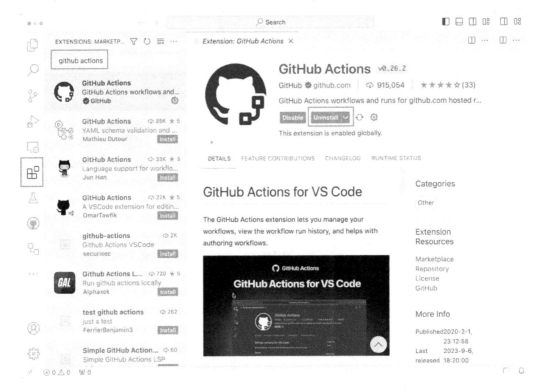

Figure 2.3 – Installing the GitHub Actions extension for VS Code

The extension provides the following features:

*   Managing workflows and monitoring workflow runs
*   Manually triggering workflows
*   Syntax highlighting for workflows and expressions
*   Integrated documentation
*   Validation and code completion
*   Smart validation

Smart validation is especially of great help. It supports code completion for referenced actions and reusable workflows, will parse parameters, inputs, and outputs for referenced actions, and provides validation, code completion, and inline documentation.

2. Open your locally cloned repository. There are multiple ways to do this. You could type `code .` (`.` stands for the current folder) on the command line inside the locally cloned repository folder. This will open a new instance of VS Code with the current folder open. Or use **File | Open Folder** in VS Code and select the folder in which you cloned the repository.

3. Click the GitHub Actions icon on the left side (see *Figure 2.4*) and inspect the **Current Branch** window. You can see all workflow runs in the current branch. Each run has the name of the workflow and the run ID with the hash. You can expand the workflow runs to see the jobs and steps. You can open the job log directly in VS Code, or you can open the step log in your browser:

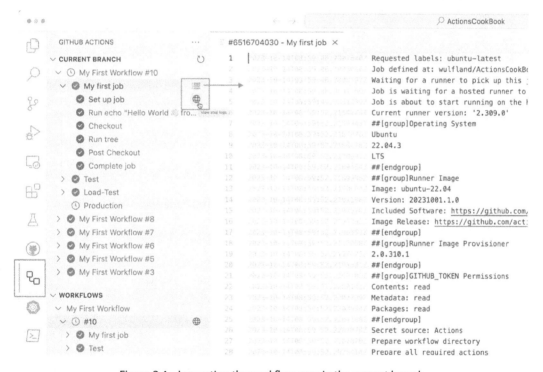

Figure 2.4 – Inspecting the workflow runs in the current branch

4. The **WORKFLOWS** window shows all the workflows in the repository in the main branch. You can pin workflows to the top of the list, open the workflow file for editing, or run a trigger for the workflow if it has a `workflow_dispatch` trigger (see *Figure 2.5*):

Figure 2.5 – The workflow window allows you to trigger and open workflows

Open the workflow file.

5. Next, play around with autocomplete in the workflow file. Remove the value behind `vars` in line 15. Note the extension knows all your variables and secrets (see *Figure 2.6*):

```
    - run: |
        echo "Hello ${{ vars. }} from ${{ github.actor }}."
        echo "My secret is  [⊘] WHO_TO_GREET
        echo "Current branch is '${{ github.ref }}'."
    - name: Checkout
      uses: actions/checkout@v4.1.0
    - run: tree
```

Figure 2.6 – Autocomplete knows all your secrets and variables

It also has the complete workflow syntax. You can press *[Control]+[Space]* at any point in the workflow file to get a list of valid elements.

VS Code detects problems in your workflow file and provides quick fixes for it (see *Figure 2.7*):

```
 MyFirstWorkflow.yml 1, M  X

.github > workflows >  MyFirstWorkflow.yml
    1    # A first workflow to play around with the editor
    2    name
    3    on:        Unexpected value 'branch'
    4      pu   View Problem    Quick Fix...
    5    branch:
    6        - main
    7    workflow_dispatch:
    8

PROBLEMS  1    OUTPUT    DEBUG CONSOLE    TERMINAL    PORTS

⌄  MyFirstWorkflow.yml  .github/workflows  1
    ⊗ Unexpected value 'branch' [Ln 5, Col 5]
```

Figure 2.7 – VS Code provides quick fixes for problems

6. In the **SETTINGS** window, you can find all the environments, secrets, and variables from your repository. You cannot create new environments, but you can add secrets and variables at the environment or repository level and edit or delete them (see *Figure 2.8*):

Figure 2.8 – Manage environments, secrets, and variables from within VS Code

How it works...

With the GitHub Actions Extension, VS Code is the perfect editor to write and execute workflows as you have everything in one place, you can work offline, and you have advanced syntax highlighting and auto-complete for workflows and expressions. That's why we are going to use it in the rest of the book to write our workflows.

There's more...

VS Code is not just available to install locally—you can also use it directly in the browser. In any GitHub repository, just press the dot [.] key to open the current repository in VS Code directly in your browser or press *[Shift]+[>]* to open it in a new tab. To open repositories directly in VS Code, you can also navigate to `https://github.dev/<owner>/<repository>`.

In **GitHub.dev**, you can work with files and Git as you do locally: first, commit your changes and then push them to GitHub. You can install extensions and you can sync your VS Code settings using your GitHub account.

However, if you need a terminal or install some frameworks, you have to do it locally or use **GitHub Codespaces** (`https://github.com/features/codespaces`). Codespaces provide you with a full remote development environment running in Microsoft Azure. You have 120 hours and 15 GB storage per month for free (180 minutes and 20 GB with the GitHub Pro Plan); after that, you pay by minute and GB. A GB costs $0.07 cent per month, and the computing for the machine ranges between $0.18 per hour for a two-core machine to $2.88 per hour for a 32-core machine. For the purposes of this book, I opted for the local version so as to not burn through the free minutes, but if you have never tried Codespaces, I encourage you to do so. It's a great way to have specialized development environments for each project you are working on.

Developing workflows in branches

Starting in a greenfield repository, it is best to create your workflows on the `main` branch. However, if you must create the workflow in an active repository that developers are working in and you don't want to get in their way, then it is possible to write workflows in a branch and merge them back to the main branch using a pull request.

However, some triggers might not work as expected. If you want to run your workflow manually using the `workflow_dispatch` trigger, your first action must be to merge the workflow with the trigger back to `main` or use the API to trigger the workflow. After that, you can author the workflow in a branch and select the branch when triggering the workflow through the UI.

If your workflow needs webhook triggers, such as `push`, `pull_request`, or `pull_request_target`, it might be necessary to create the workflow in a fork of the repository, depending on what you plan on doing with the triggers. This way, you can test and debug the workflow without interfering with the developers work, and once you are done, you can merge it back to the original repository.

Getting ready...

If you still have local changes after playing around with the workflow in the previous recipe, be sure to undo all the changes to have a clean version of the repository. You can do this by executing the following command:

```
$ git reset --hard HEAD
```

You can also do this in the Git window of VS Code (see *Figure 2.9*):

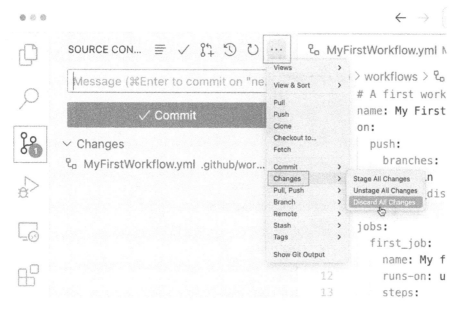

Figure 2.9 – Discard your local changes

How to do it...

1. In VS Code, click **main** in the left bottom corner, select **+ Create new branch…** in the command palette, enter new-workflow as the name for the new branch, and hit *[Enter]* (see *Figure 2.10*):

Figure 2.10 – Creating a new branch in VS Code

Alternatively, you can also use the following command:

```
$ git switch -c new-workflow
```

2. Create a new workflow file in VS Code in the **EXPLORER** window. Locate and mark the
    `.github/workflow` folder and click the **New file...** icon. Enter `DevelopInBranch.`
    `yml` as the filename and click enter (see *Figure 2.11*):

Figure 2.11 – Creating a new workflow file in VS Code

Note that VS Code automatically detects that this is a workflow file. Create a simple workflow
with a `pull_request` and `workflow_dispatch` trigger that outputs some context values
to the console, as is shown in *Listing 2.1*:

Listing 2.1 – Workflow created in a branch

```yaml
# workflow to show how to develop workflows in branches
name: Develop in a branch

on: [pull_request, workflow_dispatch]

jobs:
  job1:
    runs-on: ubuntu-latest
    steps:
      - run: |
          echo "Workflow triggered in branch '${{ github.ref
}}'."
```

```
            Echo "Workflow triggered by event '${{ github.event_
    name }}'."
            Echo "Workflow triggered by actor '${{ github.actor
    }}''."
```

3. Add the new file (**Stage Changes**), enter a commit message, and commit the changes (see *Figure 2.12*):

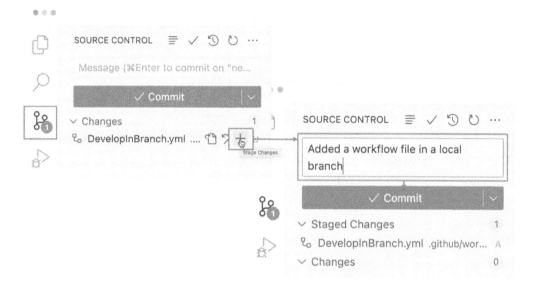

Figure 2.12 – Committing the new file in VS Code

You can also use the command line if you prefer:

```
$ git add .
$ git commit -m "Added a workflow file in local branch"
```

4. In VS Code, you can push your changes directly by clicking **Publish Branch**. From the command line, you can use the following:

```
$ git push -u origin new-workflow
```

5. Next, we'll create a pull request for our new branch. As we use the `pull_request` trigger, this will automatically run our new workflow. Go to your repository in the browser and navigate to **Pull requests**. Git will detect that you have pushed a new branch and will offer you the option to create a pull request (**Compare & pull request**, see *Figure 2.13*):

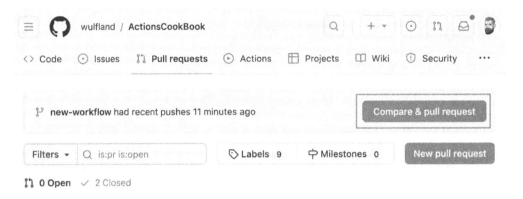

Figure 2.13 – Creating a new pull request in the browser

Just leave the default title (the commit message you added earlier) and click **Create pull request** (see *Figure 2.14*):

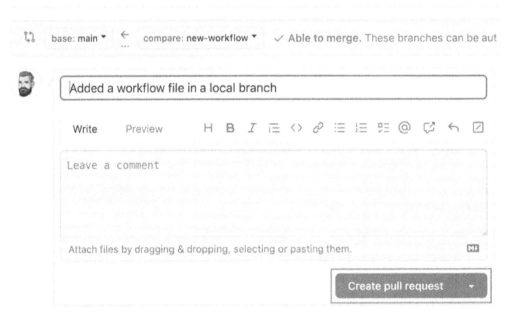

Figure 2.14 – Creating a pull request with title and description

You can also create the pull request using the GitHub CLI:

```
$ gh pr create --fill
```

The GitHub CLI

We will use the GitHub CLI (`https://cli.github.com/`) a lot throughout the book. It is available for all platforms and with a lot of package managers (Homebrew, WinGet, RPM, and many more). See `https://github.com/cli/cli#installation` for more installation instructions. After installation, you have to authenticate using `gh auth login` (see `https://cli.github.com/manual/gh_auth_login`).

---

6. Open your pull request and note that it has executed the workflow as a check automatically (see *Figure 2.15*):

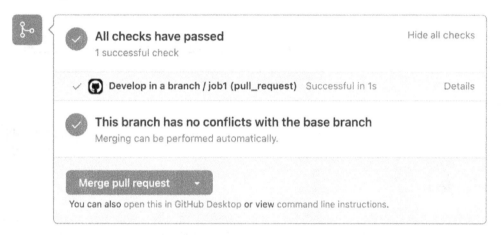

Figure 2.15 – The workflow will automatically run because of the pull_request trigger

You can also see the workflow in the **Actions** tab, but you cannot run the workflow manually. Even with the `workflow_dispatch` trigger, the button to run the workflow will only be available if you merge the workflow with that trigger to `main` once. After that, you will also be able to run it manually on your branch.

You can run the workflow in VS Code, though. Open the **GitHub Actions** extension and refresh the **Workflows** window if necessary by using the refresh icon in the top right corner. You should now see the new workflow, and you can trigger it manually using the arrow button (see *Figure 2.16*):

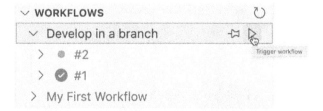

Figure 2.16 – Run the workflow manually in VS Code

How it works...

There are some limitations when it comes to some triggers, but in general, it works really well to develop your workflows in a separate branch and collaborate on the changes using pull requests.

There's more...

To take this one step further, we'll add a linter to the workflow that will be able to spot errors, security issues, and missing best practices in all workflows in your repository.

Linting workflows

In this recipe, we'll add a linting action that will check the workflow and give feedback directly in the pull request.

Getting ready...

Open the workflow file in the branch that you have for the open pull request. Do not merge the changes yet.

How to do it...

1. Go to the marketplace and search for the `actionlint`. The action we are looking for is from `devops-actions`. The action needs to access the workflow files, meaning you have to check out the repository using the checkout action first. Add the following two steps to the end of the job:

    ```
    - uses: actions/checkout@v4.1.0
    - uses: devops-actions/actionlint@v0.1.2
    ```

2. As we want the action to annotate errors in pull requests, we have to give the workflow write access to pull requests. I'll explain later how this works. For now, just add a section permissions to the job like this:

    ```
    jobs:
      job1:
        runs-on: ubuntu-latest
        permissions:
          contents: read
          pull-requests: write
    ```

3. Commit the changes to the branch `new-workflow` and push them to the remote. This will trigger the build, and the workflow should complete without errors.

4. To test the linting process, we are going to add some malicious code. If you use user-controlled input in a run event, such as the title of a pull request, attackers might be able to exploit this by injecting a script in the title. This is especially critical if you have a public repository and people can create pull requests from forks. Let's assume you output the title of a pull request using echo:

```
- run: echo "${{ github.event.pull_request.title }}"
```

Creating a pull request with the title `"Hi";ls $GITHUB_WORKSPACE;echo "-"` would execute the following command:

```
$ echo "Hi";ls $GITHUB_WORKSPACE;echo "-"
```

The script `ls $GITHUB_WORKSPACE` will be executed without errors, and from there, you can find other ways to inject more harmful script.

Add the following line to the steps in the job:

```
steps:
  - run: echo "PR title is '${{ github.event.pull_request.title
}}'."
```

5. Commit the changes to the branch `new-workflow` and push them to the remote. This time, the workflow should fail, and the checks of the pull request show an error, as is shown in *Figure 2.17*:

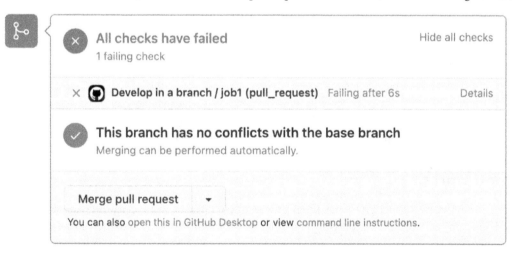

Figure 2.17 – Linting can fail a pull request if there are potential script attacks in your workflow

6. In the pull request, navigate to **Files changes**. Note that the linting action has annotated the pull request at the correct line number that contains the potential script injection attack (see *Figure 2.18*):

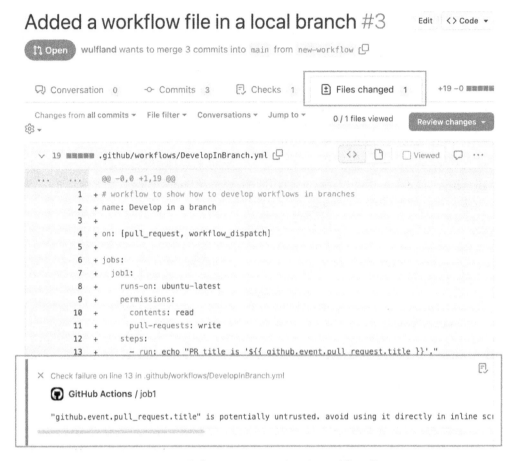

Figure 2.18 – Pull request annotations in workflow files

How it works...

The action devops-actions/actionlint (https://github.com/marketplace/actions/rhysd-actionlint) is a wrapper for @rhysd (this is the GitHub user) actionlint container (see https://github.com/rhysd/actionlint). You can run actionlint locally or on the web. It has a ton of checks that it performs on a workflow (see https://github.com/rhysd/actionlint/blob/main/docs/checks.md for a complete list). The action is a wrapper that runs action lint against all workflows found in your repository. Therefore, you have to check out the repository first. The action then uses **Problem matchers** (see https://github.com/actions/toolkit/blob/main/docs/problem-matchers.md) to annotate your workflows in pull requests. Problem matchers use regular expression patterns to read findings from a result file and annotate your pull requests with them. They are activated and deactivated by the workflow commands add-matcher and remove-matcher. Workflow commands can be used in workflow steps and actions to communicate with the workflow and the runner machine. They can be used to write messages to the workflow log, pass values to other steps or actions, set environment variables, or write debug messages.

Workflow commands use the `echo` command with a specific format:

```
echo "::workflow-command param1={data},param2={data}::{command value}"
```

If you are using JavaScript, the toolkit (`https://github.com/actions/toolkit`) provides a lot of wrappers that can be used instead of using echo to write to standard output. In the subsequent sections, you will learn some examples of useful workflow commands to write to the workflow log and to annotate files.

The problem matchers are added using the following workflow command:

```
echo "::add-matcher::$GITHUB_ACTION_PATH/actionlint-matcher.json"
```

The command takes the path to the results file. You could disable matching using `remove-matcher` and passing in the owner:

```
echo "::remove-matcher owner= actionlint::"
```

To access the pull request, the workflow uses the `GITHUB_TOKEN`, which is a special token that is automatically created by GitHub and can be accessed through the `github` context (`github.token`) or the secrets context (`secrets.GITHUB_TOKEN`). The token can be accessed by a GitHub Action, even if the workflow does not provide it as an input or environment variable. The token can be used to authenticate the workflow when accessing GitHub resources. The default permissions can be set to permissive (read and write) or restricted (read-only), but the permissions can be adjusted in the workflow. You can see the workflow permissions in the workflow log under **Set up job | GITHUB_TOKEN Permissions**. It is best practice to always explicitly set the permissions your workflow needs. All other permissions will be set to `none` automatically. The permissions can be set for an individual job or the entire workflow.

In our case, we gave the workflow job permission to read content and write to pull requests:

```
permissions:
  contents: read
  pull-requests: write
```

There's more...

In this recipe, we added the workflow linter as a check for the pull request by just adding the `pull_request` trigger to the workflow. However, we would still be able to merge the changes back to the `main` branch, even if the check fails. To prevent workflows with linting errors from being merged, you can enable **branch protection** (see `https://docs.github.com/en/`

`repositories/configuring-branches-and-merges-in-your-repository/` `managing-protected-branches/about-protected-branches`) or create **rulesets** (see `https://docs.github.com/en/repositories/configuring-branches-and-` `merges-in-your-repository/managing-rulesets/about-rulesets`). Together with **codeowners** (see `https://docs.github.com/en/repositories/managing-` `your-repositorys-settings-and-features/customizing-your-repository/` `about-code-owners`), you can ensure that only workflows without linting errors that have been manually reviewed by a team or person are merged back to your main branch.

Writing messages to the log

What problem matchers do based on existing result files can also be achieved by writing individual warning or error events to the log by also using workflow commands. In this recipe, we will add some output to our workflow and annotate our workflow file.

Getting ready...

Make sure you still have your pull request from the previous recipe open. Just use VS Code to add additional changes, and pushing will automatically trigger the workflow.

How to do it...

1. Open `.github/workflows/DevelopInBranch.yml` in the `new-workflow` branch in VS Code and add the following code snipped directly before the checkout action:

   ```
   - run: |
       echo "::debug::This is a debug message."
       echo "::notice::This is a notice message."
       echo "::warning::This is a warning message."
       echo "::error::This is an error message."
   ```

   This will write different kinds of messages to the workflow log and workflow summary.

2. Commit and push the changes. This will automatically trigger a new workflow run.

3. Open the workflow log and inspect the output. It should look like *Figure 2.19*:

```
✓   Run echo "::debug::This is a debug message."

1   ▼Run echo "::debug::This is a debug message."
2    echo "::debug::This is a debug message."
3    echo "::notice::This is a notice message."
4    echo "::warning::This is a warning message."
5    echo "::error::This is an error message."
6    shell: /usr/bin/bash -e {0}
7   Notice: This is a notice message.
8   Warning: This is a warning message.
9   Error: This is an error message.
```

Figure 2.19 – Writing messages to the workflow log

Note that the debug message is not visible. Check the workflow summary and that it contains the messages there together with our error from the linting (the result should look like *Figure 2.20*):

Figure 2.20 – Workflow annotations in the summary

4. To see the debug message in action, we can rerun the workflow job with debug logging enabled. In the workflow summary, click **Re-run jobs | Re-run all jobs**, select **Enable debug logging**, and click **Re-run jobs** (as shown in *Figure 2.21*):

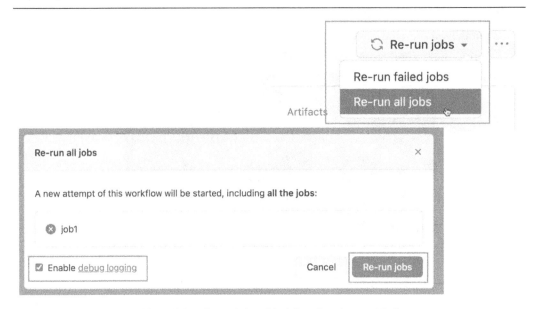

Figure 2.21 – Rerun jobs with debug logging enabled

5.   Check the workflow log again and see the additional messages (as shown in *Figure 2.22*):

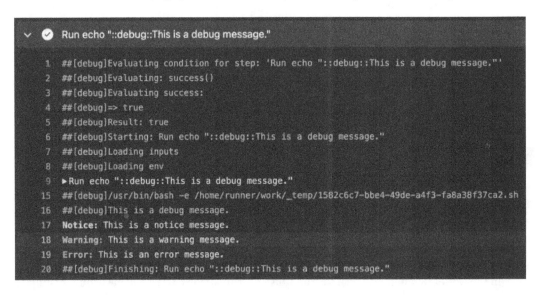

Figure 2.22 – The workflow log with debug logging on

6. However, **notice**, **warning**, and **error** can do more than just write to the log. We can use them to annotate files. Add the following snippet to the workflow:

```
- run: |
    echo "::notice file=.github/workflows/DevelopInBranch.
    yml,line=19,col=11,endColumn=51::There is a debug message that
    is not always visible!"
    echo "::warning file=.github/workflows/DevelopInBranch.
    yml,line=19,endline=21::A lot of messages"
    echo "::error title=Script Injection,file=.github/workflows/
    DevelopInBranch.yml,line=13,col=37,endColumn=68::Potential
    script injection"
```

This will add a notice annotation to line 19, a warning to lines 19 to 21, and an error to line 13 for columns 37 to 68. Adjust the values if your line numbers and indentations are different!

7. Commit and push the changes. Open the pull request and see the annotations within the **Files changed** tab (see *Figure 2.23*):

Figure 2.23 – Workflow annotation in the pull request changes

How it works...

In the same way that the matchers work, you can create warning and error messages and print them to the log. The messages will create an annotation, which can associate the message with a particular file in your repository. Optionally, your message can specify a position within the file:

```
::notice file={name},line={line},endLine={el},title={title}::{message}
::warning
file={name},line={line},endLine={el},title={title}::{message}
::error file={name},line={line},endLine={el},title={title}::{message}
```

The parameters are the following:

- **Title**: A custom title for the message
- **File**: The filename that raised the error or warning
- **Col**: Column/character number, starting at 1
- **EndColumn**: The end column number
- **Line**: The line number in the file starting with 1
- **EndLine**: The end line number

The only message that cannot annotate files is the debug message. This workflow command only accepts the message as a parameter.

Enabling debug logging

In the previous recipe, we saw that you can rerun failed jobs or all jobs with debug logging enabled. However, you can also enable or disable debug logging on a repository base.

How to do it...

We can enable or disable debug logging by adding a variable called ACTIONS_STEP_DEBUG to our repository and setting the value to true or false. This will add a very verbose output to our workflow logs and all debug messages, and this will be displayed from all actions.

You can configure the variable using the web, the GitHub CLI, or VS Code. To set the variable using the web, in the repository, navigate to **Settings** | **Secrets and variables** | **Actions** and pick the **Variables** tab (/settings/variables/actions). Click **New repository variable** (which will redirect you to /settings/variables/actions/new), enter ACTIONS_STEP_DEBUG as the name, true as the value, and click **Add variable**.

To set it using the CLI, just execute the following line:

```
$ gh variable set ACTIONS_STEP_DEBUG --body true
```

If you want to set the variable in VS Code, just open the Actions extension, navigate to **Variables |
Repository Variables**, click the + symbol (see *Figure 2.24*), enter `ACTIONS_STEP_DEBUG`, and
hit *[Enter]*; enter `true` and hit *[Enter]* again. In VS Code, it is also very convenient to change the
variable using the update option:

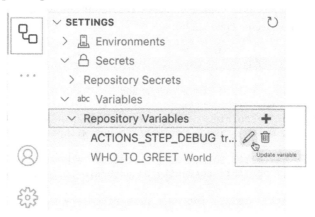

Figure 2.24 – Setting step debugging to true in VS Code

Run your workflow and inspect the verbose output.

There's more...

You can also activate additional logs for runners by setting the variable `ACTIONS_RUNNER_DEBUG`
to `true`. The runner debug log will be included in the log archive from the workflow that you can
download from the workflow job log. If you want to learn more about monitoring and troubleshooting,
then you can refer to `https://docs.github.com/en/actions/monitoring-and-
troubleshooting-workflows/enabling-debug-logging`.

Running your workflows locally

Committing workflows every time and running them on the server can be a slow process, especially for
complex workflows. In this recipe, we will learn how to run a workflow locally using **act** (`https://
github.com/nektos/act`).

Getting ready...

Act depends on Docker to run workflows. Make sure you have Docker running.

You can install act using different package managers (see `https://github.com/nektos/
act#installation-through-package-managers`). Just pick the one that fits your
environment and follow the instructions.

When running act for the first time, it will ask you to choose a Docker image to be used as the default. It will save that information to ~/.actrc. There are different images available. There are small images available (node:16-buster-slim) that will only support **NodeJS** and nothing more. The big images are more than 18 GB in size. Keep that in mind. However, with today's disk space and internet, you will get the best results by using big images. For the current workflow to run, you should pick at least the **Medium** image (see *Figure 2.25*):

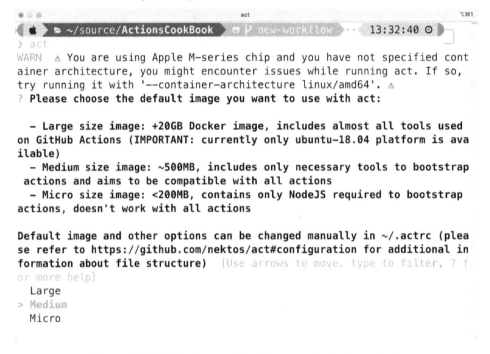

Figure 2.25 – Picking the container image on the first run of act

How to do it...

Open the command prompt in our current repository and see the new workflow branch. Act uses the workflow triggers to act on workflows, and it will default to push. As our workflow only has the pull_request trigger, you have to specify. With the -l option, act will list all workflows and jobs with the corresponding trigger in the repository. Execute the following command:

```
$ act pull_request -l
```

Inspect the workflow and the jobs. To perform a dry run, you can use the option -n:

```
$ act pull_request -n
```

Note that the workflow finishes successfully, as the linting of the workflow is not performed. To really execute the workflow in the container, run the following command:

```
$ act pull_request
```

The workflow will execute and fail in the linting step in the same way that it does in the pull request on the server. The result should look like *Figure 2.26*:

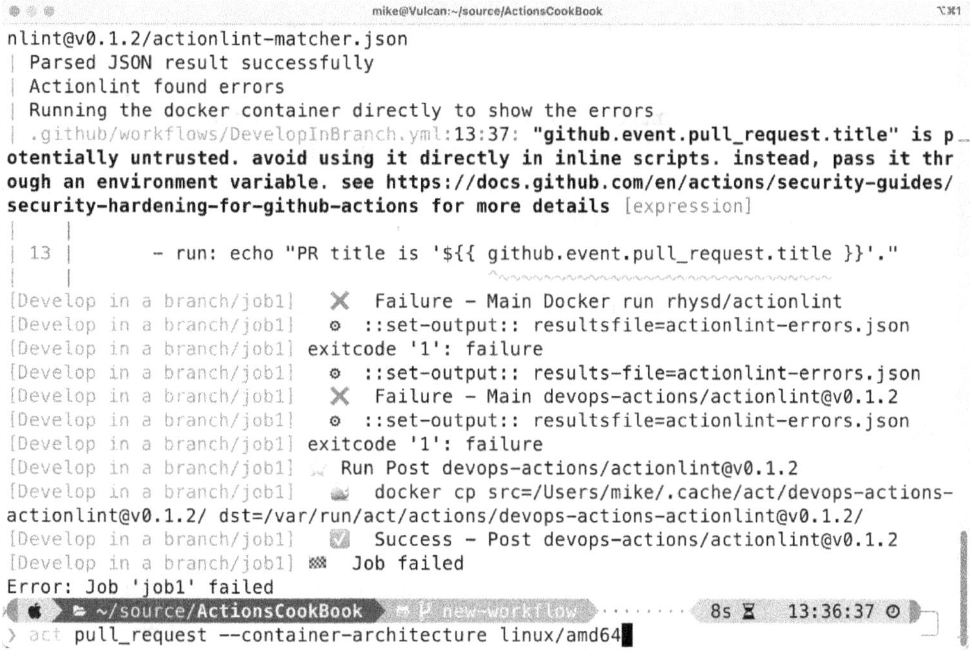

Figure 2.26 – Workflow linting fails with the same error as on the server

I think it is obvious how powerful it is to be able to run your workflows locally before pushing the changes to the server. However, depending on your workflow, the results might not be 100% reliable, and you might need to use a large Docker image that are over 20 GB in size.

How it works...

Act uses docker containers to run your workflows locally. It reads your GitHub Actions from .github/ workflows/ and determines the set of actions that need to be run. It uses the Docker API to either pull or build the necessary images, as defined in your workflow files, and finally, it determines the execution path based on the dependencies that were defined. Once it has the execution path, it then uses the Docker API to run containers for each action based on the images prepared earlier. The environment variables and filesystem are all configured to match what GitHub provides.

If your workflow uses the `GITHUB_TOKEN`, then you have to provide a **personal access token** (**PAT**); act will use it to communicate with GitHub:

```
$ act -s GITHUB_TOKEN=[insert personal access token]
```

You can use the GitHub CLI with `gh auth token` to automatically pass the token from the CLI to act:

```
$ act -s GITHUB_TOKEN="$(gh auth token)"
```

There's more...

The problem with act is that the images of the default GitHub-hosted runners are huge. For good local performance, it is just impossible to include all tools that are installed on these runners. For 90% of the workflows, this is also not necessary, as actions run in NodeJS or bring their own containers. However, especially with command line tools in custom scripts in `run:` steps, this is a problem.

What works great with act is using custom images for your workflow jobs. Instead of relying on the tools of the GitHub hosted runners, you can assign custom Docker images for jobs like this:

```
jobs:
  container-test-job:
    runs-on: ubuntu-latest
    container:
      image: custom-image:latest
```

This way, local execution and the execution on the server are basically the same. This is also a good option if you have to keep your build environments for a longer period. You can learn more about running jobs in containers here: `https://docs.github.com/en/actions/using-jobs/running-jobs-in-a-container`.

3
Building GitHub Actions

Now that you've learned how to author workflows and already used some actions from the marketplace, it is time to fully understand what actions are and how they work. In this chapter, I will explain the different types of actions and we will cover the following recipes so that you know how to write actions yourself:

- Creating a Docker container action
- Adding output parameters and using job summaries
- Creating a TypeScript action
- Creating a composite action
- Sharing actions to the marketplace
- Best practices for developing custom actions

You will learn how to pass in parameters to actions and use output parameters in subsequent workflow steps. You will also learn how to write to the workflow log and annotate changes in files from within the action and how to create rich job summaries.

Technical requirements

In the following recipes, I'll be using VS Code. You'll need a version of that and a local git client to follow along.

If you want to run the Docker image that we provision as an action locally, you will also need Docker installed on your machine. You could also use GitHub Codespaces if you want.

For the TypeScript action, you'll need to have a reasonably modern version of Node.js installed. If you are using a version manager such as nodenv or nvm, you can run nodenv install in the root of your repository to install the version specified in package.json. Otherwise, 20.x or later should work. Check the README.md file at https://github.com/actions/typescript-

`action` for updated requirements. If you don't want to install Node.js, just use GitHub Codespaces for that recipe.

Creating a Docker container action

In this recipe, you will create a simple **Docker container action** from a Dockerfile and use it in a **continuous integration** (**CI**) workflow that will run the action from within the workflow every time you change something.

Getting ready...

Create a new repository called `DockerActionRecipe`. Make it public so that you don't consume any action minutes and initialize it with a README file (see *Figure 3.1*):

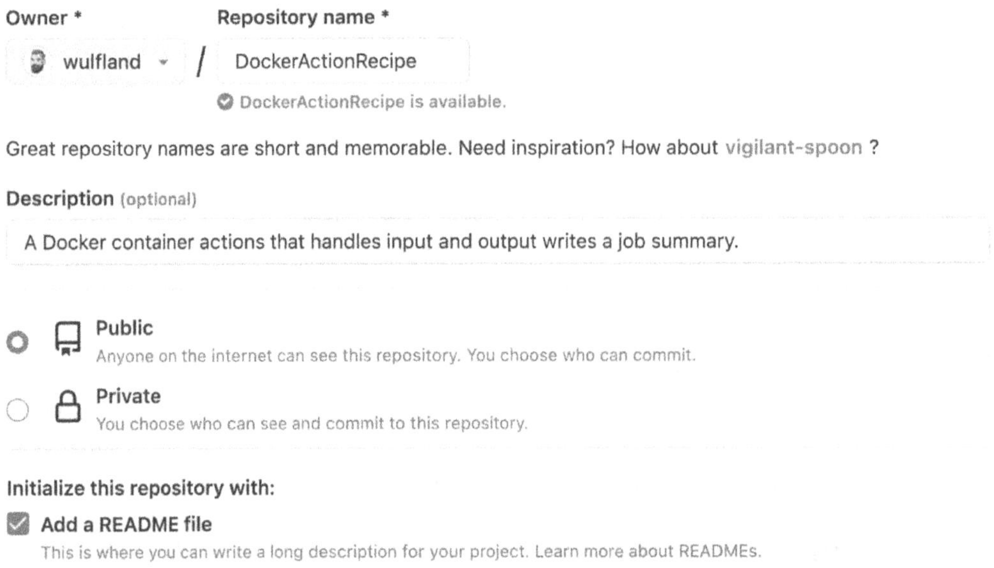

Owner *   Repository name *

  wulfland ▾   /   DockerActionRecipe

                    ✓ DockerActionRecipe is available.

Great repository names are short and memorable. Need inspiration? How about vigilant-spoon ?

Description (optional)

  A Docker container actions that handles input and output writes a job summary.

○ ▭ **Public**
        Anyone on the internet can see this repository. You choose who can commit.

○ 🔒 **Private**
        You choose who can see and commit to this repository.

Initialize this repository with:
☑ **Add a README file**
   This is where you can write a long description for your project. Learn more about READMEs.

Figure 3.1 – Creating a new repository for the Docker container action

Clone the repository locally and open it in VS Code or GitHub Codespaces.

How to do it...

1. Create a new file called `Dockerfile` in the root of the repository. Add the following content to the file:

```
# Container image that runs your code
FROM alpine:latest
CMD echo "Hello World"
```

This will create an image based on the latest Alpine image and add a layer that writes "Hello World" to the console.

2. Run the Docker container locally with the following command:

```
$ docker run $(docker build -q .)
```

It will create an image (`docker build`) and run it (`docker run`). You should be able to see `Hello World` on the console.

3. To be more flexible, let's move our script code to its own file. Create a new file called `entrypoint.sh` and add the following content:

```
#!/bin/sh -l
echo "Hello World"
```

4. Now, adjust the Dockerfile so that it executes the script instead of directly writing to the console. Copy the script file to the root of the container, and then use it as the entry point:

```
FROM alpine:3.10
COPY entrypoint.sh /entrypoint.sh
RUN chmod +x entrypoint.sh
ENTRYPOINT ["/entrypoint.sh"]
```

Note that I added the `chmod +x entrypoint.sh` command to make the script executable. Otherwise, if you try to run the container locally, it will fail, with a message stating `exec: "/entrypoint.sh": permission denied`. On all Unix-based systems, you can just run `chmod +x entrypoint.sh` locally, and the attribute will be attached to the file when committing to git. On Windows, you can use git to set the file permissions:

```
$ git add entrypoint.sh
$ git update-index --chmod=+x entrypoint.sh
```

Run the Docker container again. You should see `Hello World` again – this time from the script file:

```
$ docker run $(docker build -q .)
```

5. In the action, we want to make use of an input parameter. That's why we are going to parameterize the script. Replace the word `World` with the arguments that were passed to Docker ($@ for all arguments):

```
#!/bin/sh -l
echo "Hello $@"
```

Try to run the container again locally and pass in some words. The container will print the result, as follows:

```
$ docker run $(docker build -q .) foo bar
> Hello foo bar
```

6. Next, add a new file to your repository called `action.yml` and add the following input:

```
name: 'Docker Action Recipe'
description: 'Greet someone'
inputs:
  who-to-greet:
    description: 'Who to greet'
    required: true
    default: 'World'
runs:
  using: 'docker'
  image: 'Dockerfile'
  args:
    - ${{ inputs.who-to-greet }}
```

7. At this point, the action is ready. To test it, we'll add a local workflow file called `.github/workflows/ci.yml` that will run on every push. It will download the repository and execute our action with a custom input parameter:

```
name: Action CI

on: [push]

jobs:
  ci:
    runs-on: ubuntu-latest
    steps:
      - uses: actions/checkout@v4.1.1
      - name: Run my own container action
        uses: ./
        with:
          who-to-greet: '@wulfland'
```

> **Referencing local actions**
>
> Note that we are referencing the action by the local path, `. /` – that's the reason why we have to use the checkout action first. The workflow will use the same version the workflow runs on. You could also reference the action normally by specifying `<owner>/DockerActionRecipe@ main` – the same way you would reference it from another repository.

8. Commit and push all your changes. The push trigger will automatically run your workflow and you can inspect the output of your action. It should look like what's shown in *Figure 3.2*:

Figure 3.2 – Output of your action in the workflow

Check the output of the action via the Docker daemon and the parameter that was passed to the action.

How it works...

There are three different types of actions:

- Docker container actions
- JavaScript actions
- Composite actions

Docker container actions run on Linux only, whereas **JavaScript actions** and **Composite actions** can be used on any platform.

All actions are defined by a file called `action.yml` (or `action.yaml`) that contains the metadata that defines the action. This file cannot be named differently, meaning an action must reside in its own repository or folder. The `run` section in the `action.yml` file defines what type of action it is.

Docker container actions contain all their dependencies in the container and are therefore very consistent. They allow you to develop your actions in any language – the only restriction is that it must run on Linux. Docker container actions are slower than JavaScript actions because of the time it takes to retrieve or build the image and start the container.

Docker container actions can reference an image in a container registry, such as Docker Hub or GitHub Packages, or it can build a Dockerfile at runtime that you provide with the other action files. In this case, you must specify Dockerfile as the image name in the `action.yml` file.

There's more...

Container actions are very powerful as you can write them in any language. You can return output parameters to your workflow, write messages to the workflow log, annotate files in pull requests, and write rich job summaries. In the next short recipe, we'll add output parameters and write to the job summary.

Adding output parameters and using job summaries

In this recipe, we'll add an output parameter to the action that can be used in subsequent steps, and we are going to write content to the workflow job summary.

Getting ready...

You will have to finish the previous recipe to continue with this one.

How to do it...

1. Open the `action.yml` file and add the following code right under the `inputs` section but before the `runs` section:

    ```
    outputs:
      answer:
        description: 'The answer to everything (always 42)'
    ```

    This defines one output with an ID of `answer`.

2. Next, open `entrypoint.sh` and add the following line to the end of the file:

    ```
    echo "answer=42" >> $GITHUB_OUTPUT
    ```

    This will set the output value for `answer` to `42`.

3. Now, add the following lines to the end of `entrypoint.sh` to write some Markdown and HTML to the step summary:

    ```
    echo "### Hello $@! :rocket:" >> $GITHUB_STEP_SUMMARY
    echo "<h3> The answer from Deep Thought is 42 :robot:</h3>" >>
    $GITHUB_STEP_SUMMARY
    ```

4. Before we commit the changes, we must adjust the workflow file, `.github/workflows/ci.yml`, so that it uses the output parameter. Add an ID of `my-action` to the step that executes our action, like this:

    ```
    - name: Run my own container action
      id: my-action
      uses: ./
      with:
        who-to-greet: '@wulfland'
    ```

    Add another step that outputs the result to the workflow log:

    ```
    - name: Output the answer
      run: echo "The answer is ${{ steps.my-action.outputs.answer
    }}"
    ```

5. To fail the CI build when an unexpected result is returned from the Docker container action, we must add a new step that only gets executed when the result is not expected. Returning a non-zero value (for example, `exit 1`) will indicate to the workflow that the step has failed. We can use file annotation to indicate where the error is (the same way we did in *Chapter 2*):

    ```
    - name: Test the container
      if: ${{ steps.my-action.outputs.answer != 42 }}
      run: |
        echo "::error file=entrypoint.sh,line=4,title=Error in
    container::The answer was not expected"
        exit 1
    ```

6. Commit and push all your changes. The build will run automatically and should succeed. Inspect the workflow log and ensure the output parameter was passed to the next steps correctly (see *Figure 3.3*):

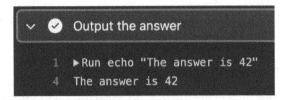

Figure 3.3 – The output of the Docker container action

Also, take a look at the job summary on the summary page, which has rendered our Markdown/
HTML (see *Figure 3.4*):

Figure 3.4 – Job summary on the workflow summary page

7. Finally, we want to ensure that our CI build fails if an unexpected value is returned. Open
    entrypoint.sh and change 42 to something different (for example, 7).

    Switch to another branch, commit and push, and create a new pull request:

    ```
    $ git switch -c fail-ci-build
    $ git commit -m "Fail CI build"
    $ git push -u origin fail-ci-build
    $ gh pr create --fill
    ```

The checks for the pull request will fail, and it will annotate the corresponding file (see *Figure 3.5*):

Figure 3.5 – The CI build will fail the pull request if the output is unexpected

Note how GitHub marks the change and annotates the exact line.

How it works...

Let's understand how the code works.

Environment files

Passing output values to subsequent steps and jobs works by piping a name-value pair to the *environment file* – that is, $GITHUB_OUTPUT:

```
echo "{name}={value}" >> "$GITHUB_OUTPUT"
```

The >> operator appends the name-value pair to the end of the file. The path and filename of the file are stored in the $GITHUB_OUTPUT environment variable. You can access the output using the output property of the step in the steps context:

```
"${{ steps.<step-id>.outputs.<name> }}"
```

Outputs are Unicode strings and cannot exceed 1 MB in size. The total of all outputs in a workflow run cannot exceed 50 MB.

Another use case for environment files is setting environment variables for subsequent steps in a job. The path to the corresponding environment file is stored in $GITHUB_ENV. You just append another name-value pair to the end of the file, like this:

```
echo "{environment_variable_name}={value}" >> "$GITHUB_ENV"
```

Note that the name is case-sensitive! Here is a complete example of how to set an environment variable in one step and access it in a subsequent step using the env context:

```
steps:
  - name: Set the value
    id: step_one
    run: |
      echo "action_state=yellow" >> "$GITHUB_ENV"

  - run: |
      echo "${{ env.action_state }}" # This will output 'yellow'
```

Job summaries

You can set custom Markdown for each job in a workflow. The rendered Markdown will be displayed on the *summary* page of the workflow run. Job summaries can be used to display content, such as test or code coverage results, so that someone viewing the result of a workflow run doesn't need to go into the logs or an external system.

Job summaries support GitHub-flavored markdown. But since Markdown is HTML, you can also output HTML to the job summary file. Note that in this recipe, my GitHub username, @wulfland, is a link to my profile with a preview and that all the GitHub emojis are supported.

Adding results from your step to the job summary can be achieved by appending Markdown to the following file:

```
echo "{markdown content}" >> $GITHUB_STEP_SUMMARY
```

The steps are isolated and restricted to 1 MiB (1.04858 MB) so that malformed Markdown from a single step cannot break Markdown rendering for subsequent steps. Only 20 steps can be written to the summary; the output of any step after that will not be visible.

In the next recipe, we will use the toolkit to write more complex things to job summaries.

For a complete reference on environment files and job summaries, please refer to https://docs.github.com/en/actions/using-workflows/workflow-commands-for-github-actions?tool=bash#environment-files.

Expressions and conditional execution

We used expressions ($\{\{ \ldots \}\}$) in *Chapters 1* and *2* to output values from context objects. In this recipe, we used an expression with the `if` property of a step to conditionally execute it:

```
- name: Test the container
  if: ${{ steps.my-action.outputs.answer != 42 }}
```

This step will only be executed *if* the value of an output `answer` of an action with an ID of `my-action` in the `steps` context does not equal `42`.

The `if` property also exists for `jobs` to conditionally execute these as well.

When you use expressions in an `if` property, you can, optionally, omit the $\{\{ \}\}$ expression syntax because GitHub Actions automatically evaluates the `if` condition as an expression.

For conditional execution, the expression must return `true` or `false`. To write expressions and compare context with static values, you can use the operators provided in *Table 3.1*:

Operator	Description
( )	Logical grouping
[ ]	Index
.	Property de-reference
!	Not
< , <=	Less than, less than or equal to
> , >=	Greater than, greater than or equal to
==	Equal
!=	Not equal
&&	And
||	Or

Table 3.1 – Operators for expressions

GitHub offers a set of built-in functions that you can use in expressions. They can help you search strings, format output, or work with arrays. See *Table 3.2* for a list of available functions:

Function	Description
`contains(search, item)`	Returns `true` if `search` contains `item`. Examples: `contains('Hello world', 'llo')` returns `true`. `contains(github.event.issue.labels.*.name, 'bug')` returns `true` if the issue related to the event has a label bug.
`startsWith(search, iten)`	Returns `true` when `search` starts with `item`.
`endsWith(search, item)`	Returns `true` when `search` ends with `item`.
`format(string, v0, v1, ...)`	Replaces values in `string`. Example: `format('Hello {0} {1} {2}', 'Mona', 'the', 'Octocat')` returns `'Hello Mona the Octocat'`.
`join(array, optS)`	All values in `array` are concatenated into a string. If you provide the optional separator, `optS`, it is inserted between the concatenated values. Otherwise, the default separator, `,`, is used.
`toJSON(value)`	Returns a pretty-print JSON representation of `value`.
`fromJSON(value)`	Returns a JSON object or JSON data type for `value`.
`hashFiles(path)`	Returns a single hash for the set of files that matches the `path` pattern.

Table 3.2 – Built-in functions in GitHub for expressions

There's more...

There are also some special functions to check the status of the current job. In the following example, the last step would only be executed if a previous step of the jobs failed – meaning it returns a non-zero value:

```
steps:
  - run: exit 1
  - name: The job has failed
    if: ${{ failure() }}
```

For a list of available functions to check the status of the job, see *Table 3.3*:

Function	Description
success()	Returns true if none of the previous steps have failed or been canceled.
always()	Returns true, even if a previous step was canceled, and causes the step to always get executed anyway.
cancelled()	Returns only true if the workflow was canceled.
failure()	Returns true if a previous step of the job had failed.

Table 3.3 – Functions to check the status of the workflow job

These functions can also help you to conditionally execute some steps and perform – for example – a cleanup. To learn more about expressions for conditional execution, please refer to https://docs. github.com/en/actions/learn-github-actions/expressions and https:// docs.github.com/en/actions/using-jobs/using-conditions-to-control- job-execution.

Creating a TypeScript action

In this recipe, you will create a basic **TypeScript action** from a template, build and publish it, and use it in a workflow.

Getting ready...

Make sure you have a reasonably modern version of Node.js (https://nodejs.org/en/ download) installed. Follow these steps:

1. Go to https://github.com/actions/typescript-action and click **Use this template** | **Create a new repository** in the top-right corner of the repository (see *Figure 3.6*):

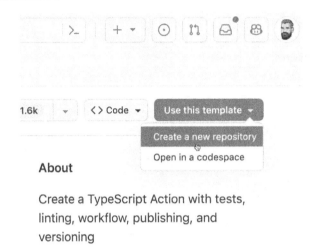

Figure 3.6 – Creating a new repository from the typescript-action template

2. Choose your GitHub account as the owner, name it `TypeScriptActionRecipe`, leave its visibility set to **Public**, and click **Create repository**.

3. Clone the repository locally and open the folder in VS Code.

How to do it...

1. Open a terminal and go to the root of the repository. Install all the necessary dependencies:

    ```
    $ npm install
    ```

    The repository contains some unit tests. Run them to check that everything is OK:

    ```
    $ npm test
    ```

2. Open the `action.yml` file and update the metadata for `name`, `description`, and `author`:

    ```
    name: 'TypeScript Action Recipe'
    description: 'Waites for some milliseconds, writes an awesome
    job summary to the workflow output and returns the current date
    and time.'
    author: 'Michael Kaufmann'
    ```

    Ignore the branding for now – but note that the action has a defined input parameter (`milliseconds`) and output parameter (`time`):

    ```
    # Define your inputs here.
    inputs:
      milliseconds:
    ```

```
      description: 'Your input description here'
      required: true
      default: '1000'

# Define your outputs here.
outputs:
  time:
      description: 'Your output description here'
```

3. Open the `src/main.ts` file and locate the `run` function:

    ```
    export async function run(): Promise<void> {
    ```

    Note that it uses the toolkit (`@actions/core`; see `https://github.com/actions/toolkit`) to read the input parameter:

    ```
    const ms: string = core.getInput('milliseconds')
    ```

    It also uses the toolkit to write debug messages. This is similar to the `echo "::debug::{debug message}"` workflow command that we used in *Chapter 2*:

    ```
    core.debug(`Waiting ${ms} milliseconds ...`)
    ```

    The same is true for setting the output parameter. This is the same as writing to the `GITHUB_OUTPUT` environment file:

    ```
    echo "answer=42" >> $GITHUB_OUTPUT
    ```

    With the toolkit, this looks as follows:

    ```
    core.setOutput('time', new Date().toTimeString())
    ```

4. The next step is to write our job summary beneath `core.setOutput` in the `run` function. Start with `core.summary` and note that you have auto-complete to help you with all the available functions and syntax. Add a h2 heading:

    ```
    // Write an advanced job summary
    core.summary
      .addHeading('Advanced Job Summary', 'h2')
    ```

    The `core.summary` object has a fluent interface. This means you can add new methods directly to the output of the previous one. Add an image and set the size to 64x64:

    ```
    .addImage(
      'https://octodex.github.com/images/droidtocat.png',
      'Droidtocat',
      {
        width: '64',
        height: '64'
      }
    )
    ```

Add a table with some data:

```
.addTable([
  [
    { data: 'File', header: true },
    { data: 'Result', header: true }
  ],
  ['foo.js', 'Pass ☑'],
  ['bar.js', 'Fail ✖'],
  ['test.js', 'Pass ☑']
])
```

And a simple link:

```
.addLink('My custom link', 'https://writeabout.net')
```

You have to finish the summary using the `write` function, to write the buffer to the environment file:

```
.write()
```

You can check the code here: `https://github.com/wulfland/ TypeScriptActionRecipe`.

5. Package the TypeScript for distribution. This is an important step, and you have to do it every time you modify a `.ts` file!

```
$ npm run bundle
```

6. Commit your changes. This will automatically trigger some workflows – and while they run, you can use the time to see what they are doing. `.github/workflows/linter.yml` runs on all push and pull requests to `main` and it will lint your code base:

```
- name: Lint Code Base
  id: super-linter
  uses: super-linter/super-linter/slim@v5
  env:
    DEFAULT_BRANCH: main
    GITHUB_TOKEN: ${{ secrets.GITHUB_TOKEN }}
    TYPESCRIPT_DEFAULT_STYLE: prettier
    VALIDATE_JSCPD: false
```

If this fails, then you probably forgot to run `npm run bundle` before committing your changes. You can also run `prettier` locally to make sure you adhere to the linting standards:

```
$ npx prettier . --check
$ npx prettier . --write
```

The `.github/workflows/codeql-analysis.yml` workflow will also run on every push and pull request to `main` – but also once a week. It will scan your code for security vulnerabilities.

Check dist (`.github/workflows/check-dist.yml`) is a simple workflow that will run `npm run bundle` and compare the output with your `dist` folder. It will fail if you forget to package your TypeScript using a small and simple script:

```
- name: Compare Expected and Actual Directories
  id: diff
  run: |
    if [ "$(git diff --ignore-space-at-eol --text dist/ | wc
-1)" -gt "0" ]; then
      echo "Detected uncommitted changes after build. See status
below:"
      git diff --ignore-space-at-eol --text dist/
      exit 1
    fi
```

The last one is the **CI build** (`.github/workflows/ci.yml`). It has two jobs. One installs all your dependencies and runs the unit tests, while the other executes your action and uses the output, as we did in *Chapter 2*:

```
- name: Test Local Action
  id: test-action
  uses: ./
  with:
    milliseconds: 1000

- name: Print Output
  id: output
  run: echo "${{ steps.test-action.outputs.time }}"
```

The `test-typescript` job will fail as we did not adjust the unit tests, but the second job should succeed. At this point, you can inspect your job summary; it should look like what's shown in *Figure 3.6*:

Figure 3.7 – The job summary created using the toolkit

Also, inspect the value of the output parameter in your workflow log (see *Figure 3.7*):

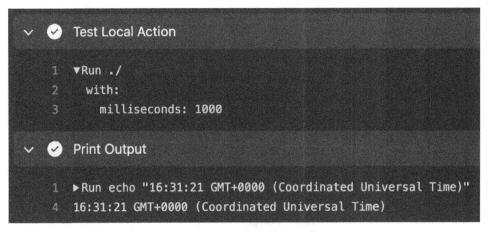

Figure 3.8 – Output of the TypeScript action in the workflow log

---

Fixing the unit tests

I haven't included a recipe for adjusting the unit tests in this book as this book is about GitHub Actions and not about TypeScript. But if you want to fix the tests, you can start by looking at the example at `https://github.com/wulfland/TypeScriptActionRecipe/blob/main/__tests__/main.test.ts`.

7. If you go back to `main.ts`, at the end, you will see that the action will fail if it encounters an error. It does so by using `core.setFailed` instead of returning a non-zero value:

```
} catch (error) {
    // Fail the workflow run if an error occurs
    if (error instanceof Error) core.setFailed(error.message)
}
```

8. Create a new branch called `fail-ci-build` and switch to it:

```
$ git switch -c fail-ci-build
```

Add the following code block to the catch block before the line that contains `core.setFailed`:

```
core.error('Something bad happened', {
    title: 'Bad Error',
    file: '.github/workflows/ci.yml',
    startLine: 59,
    startColumn: 11,
    endColumn: 23
})
```

In `.github/workflows/ci.yml`, modify the `milliseconds` argument to something that cannot be converted into an integer:

```
- name: Test Local Action
  id: test-action
  uses: ./
  with:
      milliseconds: xxx
```

Package the TypeScript, commit your changes, and create a pull request:

```
$ npm run bundle
$ git add .
$ git commit -m "Fail CI build"
$ git push –set-upstream origin fail-ci-build
$ gh pr create --fill
```

9.   Inspect the pull request you created and note the output in the log from `core.setFailed` and `core.error` (see *Figure 3.9*):

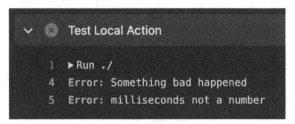

Figure 3.9 – Output of the failed action in the workflow log

Also, note that the annotation is displayed in the workflow file (see *Figure 3.10*):

```
v  ↕ 2 ■■□□□ .github/workflows/ci.yml ⬚

    ↑                @@ -56,7 +56,7 @@
56    56                    id: test-action
57    57                    uses: ./
58    58                    with:
59          -                   milliseconds: 1000
      59    +                   milliseconds: xxx

X  Check failure on line 59 in .github/workflows/ci.yml

   🐙 GitHub Actions / GitHub Actions Test

   Bad Error

   Something bad happened

60    60
61    61                - name: Print Output
62    62                  id: output
    ↓
```

Figure 3.10 – Annotations from the toolkit

The tooling around TypeScript and the toolkit make developing GitHub actions comfortable and are a great help in authoring high-quality actions.

How it works...

TypeScript actions are not so different from container actions – they just run in a Node.js environment instead of a Docker container. TypeScript is just a layer on top of JavaScript that gets transpiled to JavaScript (into the `/dist` folder) if you run `npm run bundle`. If you are new to TypeScript, all these tools might seem overwhelming – but these tools also make it quite easy to get started. Auto-complete and IntelliSense in VS Code, automated linting and formatting, unit testing with mocking – there are a lot of tools that will help you write good code.

There's more...

We just have scratched the surface of what you can do with the toolkit (`https://github.com/actions/toolkit`). It can also help you work with the GitHub REST or GraphQL API, OICD tokens, and many more. If you plan to do more with GitHub Actions, it's worth learning at least a little bit of Typescript to be able to leverage the power of the toolkit.

Creating a composite action

The third type of actions, besides Docker container actions and JavaScript/TypeScript actions, are **composite actions**. Composite actions are a wrapper for other actions. In this recipe, you will create a simple composite action and use it in a workflow – once with a bash script and once with a GitHub script.

Getting ready...

Create a new repository called `CompositeActionRecipe`. Make it public so that you don't consume any action minutes and initialize it with a README file. Clone the repository locally and open it in VS Code or open it in GitHub Codespaces.

How to do it...

1. Add a new file called `action.yml` to the root of the repository. Add a name and description:

   ```
   name: 'Composite Action Recipe'
   description: 'Greets the user and returns 42.'
   ```

2. Add an input called `who-to-greet` and an output called `answer`. Note that you need the step ID to access the output. We'll add that in the next step:

   ```
   inputs:
     who-to-greet:
       description: 'Who to greet'
       required: true
       default: 'World'
   outputs:
     answer:
       description: "Answer to life, the universe, and everything"
       value: ${{ steps.deep-thought.outputs.answer }}
   ```

3. Next, add the `runs` section, set `using` to `composite`, and add one step that executes a bash script. Note that you have to specify the shell in composite actions, and you cannot rely on the default shell:

```
runs:
  using: "composite"
  steps:
    - name: Awesome bash script action
      id: deep-thought
      shell: bash
      run: |
        echo "Hello '${{ inputs.who-to-greet }}'."
        echo "answer=42" >> $GITHUB_OUTPUT
        echo "So long, and thanks for all the fish."
```

We use the input to write a greeting message to the workflow log and we set the output parameter to 42 the same way we did it in the container action. The only difference is that we execute the script directly on the workflow runner and not in a container.

4. Add a new workflow file called `.github/workflows/ci.yml` and configure it so that it runs on push and `pull_requests`:

```
name: CI Workflow
on: [push, pull_request]
```

Add a job that checks out the repository and runs the composite action with an input parameter. Give the step an ID to access the output, like this:

```
jobs:
  ci-job:
    runs-on: ubuntu-latest
    steps:
      - uses: actions/checkout@v4.1.1
      - name: Run my own composite action
        id: my-action
        uses: ./
        with:
          who-to-greet: '@wulfland'
```

Then, add an additional step that writes the output to the console:

```
      - name: Output the answer
        run: echo "The answer is ${{ steps.my-action.
outputs.answer }}"
```

These concepts should be already familiar from looking at the other action types.

5. Commit and push your changes. This will trigger the workflow; you can inspect the workflow log to see your action in action.

How it works...

Composite actions are wrappers for steps and other actions. You can use them to bundle together multiple run commands or actions – whether you own them or they're from the marketplace – and you can provide default values for other actions to the users in your organization.

Composite actions have steps in the runs section of the `action.yml` file – like you would have in a normal workflow. You can access input arguments using the `inputs` context. Output parameters can be accessed using the `outputs` context of the step in the `steps` context.

There's more...

Composite actions get executed directly on the workflow runner. Because every runner has Node.js installed, you can run JavaScript and TypeScript in a composite action as well. We will use an action called `github-script` to leverage the power of the toolkit in a composite action. Of course, you can also use the action directly in your workflows.

Using github-script in a composite action to add a comment to an issue

In this recipe, we will use an action called `github-script` to leverage the power of the toolkit inside a composite action.

How to do it...

1. Remove the run step with the bash script from the `action.yml` file and replace it with a `github-script` action:

    ```
    - name: Awesome github script action
      uses: actions/github-script@v6
      with:
        script: |
    ```

2. Use the toolkit to read the input parameter, write a greeting to the log, and set the output parameter. This should be familiar from the *Creating a TypeScript action* recipe:

    ```
    var whoToGreet = core.getInput('who-to-greet')
    core.notice(`Hello ${whoToGreet}`)
    core.setOutput('answer', 42)
    ```

3. Now, we want to add some additional functionality. If the workflow was triggered by an `issues` event, then we want to add a comment to that issue. Check that the event that triggered the workflow was in fact `issues` and that we used `github.rest.issues` to create a comment:

```
if (context.eventName === 'issues') {
  github.rest.issues.createComment({
    issue_number: context.issue.number,
    owner: context.repo.owner,
    repo: context.repo.repo,
    body: '🙂 Thanks for reporting!'
  })
}
```

4. In the CI workflow, add an additional trigger that executes the workflow when a new issue is opened:

```
on:
  issues:
    types: [opened]
```

5. Commit and push your changes.

6. Create a new issue in your repository under **Issues** | **New issue** (`issues/new`), give it a title, and save it. The workflow will run and add a comment on that issue, as shown in *Figure 3.11*:

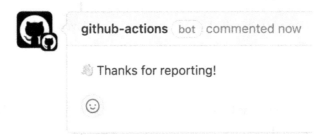

Figure 3.11 – The github-script action commenting on a new issue

The `github-script` action is a great way to prototype things using the power of the toolkit and without the need to create a separate action.

How it works...

The `github-script` action makes it easy to quickly write a script in your workflow or composite action that uses the GitHub API and the workflow run context. It uses an input named `script` that contains the body of an asynchronous function call. The following arguments will be provided by the action:

- `github`: A pre-authenticated `octokit/rest.js` client with pagination plugins

- `context`: An object containing the context of the workflow run

- `core`: A reference to the `@actions/core` package

- `glob`: A reference to the `@actions/glob` package

- `io`: A reference to the `@actions/io` package

- `exec`: A reference to the `@actions/exec` package

- `fetch`: A reference to the `node-fetch` package

- `require`: A proxy wrapper around the normal Node.js `require` to enable requiring relative paths to the current working directory and requiring npm packages to be installed in the current working directory

Since the script is just a function body, these values will already be defined, so you don't have to import them and you can directly use them as we did with `core`, `context`, and `github` in our example.

To learn more about `github-script`, please visit `https://github.com/actions/github-script`.

There's more...

The editing and debugging experience in the YAML file are not great. However, you can use the `script` file in the action like this:

```
with:
  script: |
    const script = require('./path/to/script.js')
    await script({github, context, core})
```

This way, you can combine the simplicity of the `github-script` action with a better authoring experience for JavaScript.

The `github-script` action is a great way to quickly try something out and create a proof of concept for some integrations. With composite actions, you can gradually put it into building blocks that are easy to share. Composite actions are also an easy way to package reusable functionality. As your solution evolves, you probably want to think about moving it to a TypeScript or Docker container action for better maintainability.

Sharing actions to the marketplace

The power of GitHub actions is the community – and sharing is caring. That's why the GitHub marketplace plays an essential role in empowering community-based workflows. In this recipe, you will add branding and other metadata to one of the actions and share it in the marketplace.

Getting ready...

I will use the Docker container action we created earlier for this recipe – but you can also use the TypeScript action or composite action. It doesn't matter. So long as the action resides in its own public repository, it'll work.

How to do it...

1. Navigate to the root of your repository in your browser. GitHub will detect that your repository contains an `action.yml` file and will propose that you publish a release in a blue banner (see *Figure 3.12*):

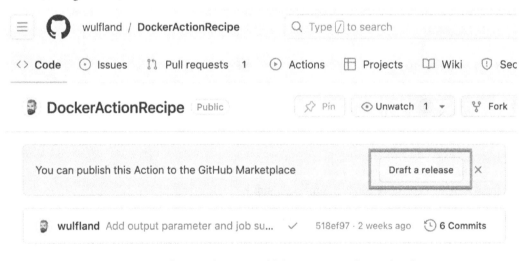

Figure 3.12 – Drafting a release to publish an action to the marketplace

This is the same as going to **Releases** and clicking **Draft a new release** (`/releases/new`).

2. In the dialogue, GitHub will show some warnings to help you improve your marketplace listing (see *Figure 3.13*):

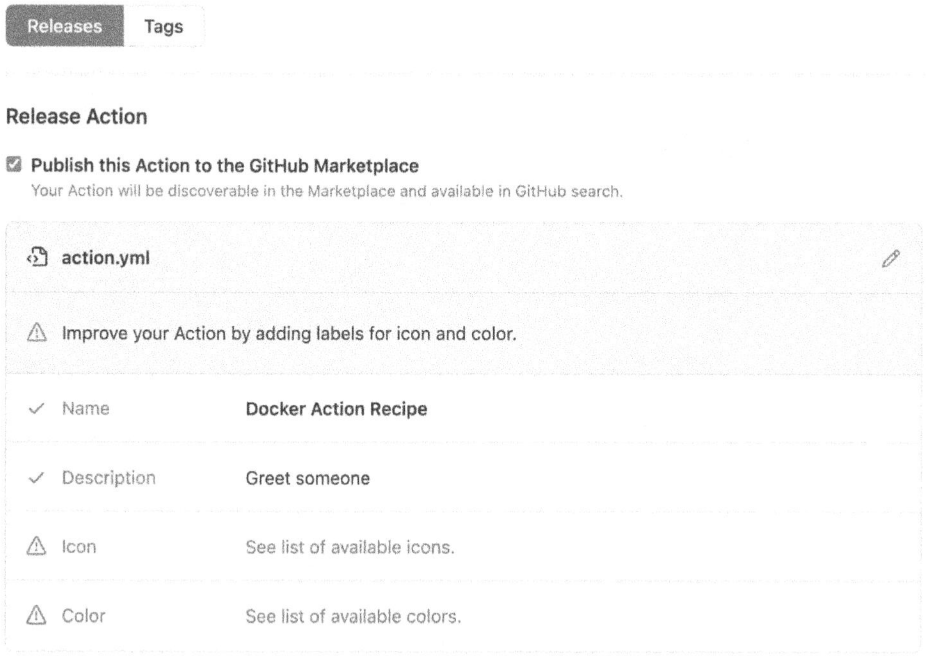

Releases   Tags

Release Action

☑ **Publish this Action to the GitHub Marketplace**
Your Action will be discoverable in the Marketplace and available in GitHub search.

action.yml	
⚠ Improve your Action by adding labels for icon and color.	
✓ Name	**Docker Action Recipe**
✓ Description	Greet someone
⚠ Icon	See list of available icons.
⚠ Color	See list of available colors.

Figure 3.13 – Guidance for publishing actions to the marketplace

Click on the link for available icons and colors and pick one of each. I'll go with `bell` and `purple`.

3. Open your `action.yml` file in VS Code and add branding and author information:

```
name: 'Docker Action Recipe'
description: 'Greet someone'
branding:
  icon: bell
  color: purple
author: 'Michael Kaufmann'
```

4. A good `README.md` file is important for the marketplace listing. Add a section for inputs, outputs, and a usage example. Check out `https://github.com/wulfland/DockerActionRecipe` for a suggestion.

5. Commit and push your changes.

6. Go back to your browser and refresh the new release window. The check should now show no warnings and look like what's shown in *Figure 3.14*):

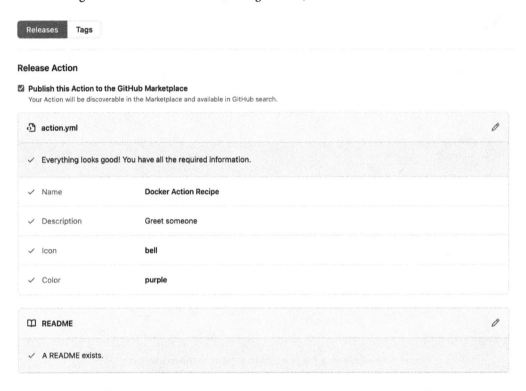

Figure 3.14 – The check is successful as has a unique name, branding, description, and a README file

7. Click **Choose a tag**, enter v1.0, and click **Create new tag** (see *Figure 3.15*):

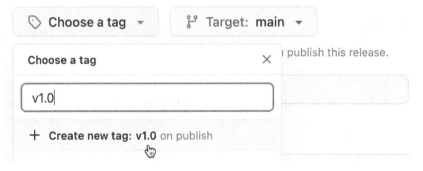

Figure 3.15 – Creating or choosing a tag to create a release

8. Add v1.0 as the title of the release. Note that you can automatically generate release notes for your release. However, the results will only be very good if you work with pull requests. You can also add a description manually.

9. Click **Publish release**. In the release, you will see the Marketplace and Latest labels (see *Figure 3.16*):

Releases / v1.0

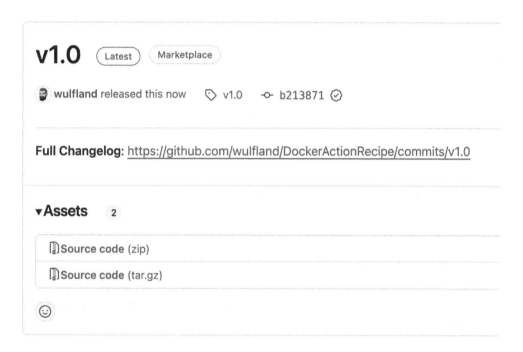

Figure 3.16 – Content of a release in your repository

10. Click the Marketplace label. This will bring you to your marketplace listing.

11. Note the icon with the color you specified under branding in your action.yml, next to the name of the action. The README.md file of your repository will cover the biggest part of the listing. On the right, you have a button to **Delist** the action (remove it from the marketplace), a version picker, and important links. The first one will bring you back to your repository (see *Figure 3.17* for the regions of the listing):

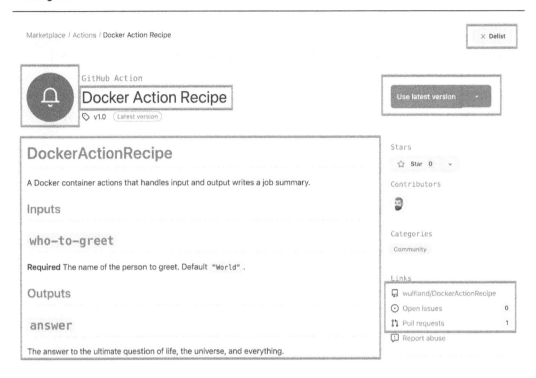

Figure 3.17 – A marketplace listing

12. Go back to your repository and create a new release called `v1.1` by repeating the same steps. Note that this time, the dialogue has a flag set as the latest release (see *Figure 3.18*). If you forget this, GitHub will not label the release as the latest – independent of the version number you pick as a label:

Figure 3.18 – You have to manually set a release as the latest when creating or editing it

13. Finally, create a new workflow file or edit an existing workflow in your browser. On the right-hand side, enter the name of your action in the marketplace window. Your action should be

found instantly. Note the installation instructions that GitHub automatically creates from your release (see *Figure 3.19*):

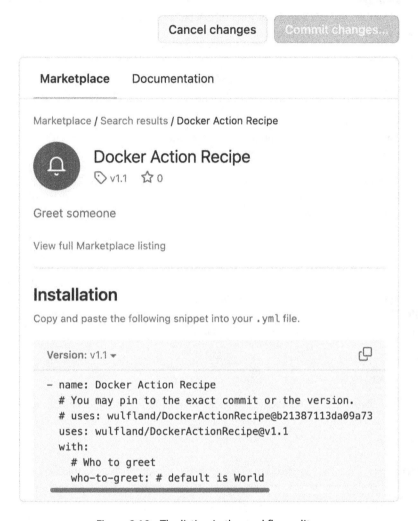

Figure 3.19 – The listing in the workflow editor

14. Delist your action if you don't want to keep it in the marketplace, as shown in *Figure 3.17*.

How it works...

The **GitHub marketplace** is built on top of **GitHub releases** (see (`https://docs.github.com/en/repositories/releasing-projects-on-github`), which are built on top of **git tags**.

A tag can be any text – but you are encouraged to use **semantic versioning** to give the versions a meaning.

Semantic versioning is a formal convention for specifying version numbers for software. It consists of different parts with different meanings. Examples of semantic version numbers are `1.0.0` or `1.5.99-beta`. The format is as follows:

```
<major>.<minor>.<patch>-<pre>
```

Let's take a closer look:

- **Major version**: A numeric identifier that gets increased if the version is not backward-compatible and has breaking changes. An update to a new major version must be handled with caution! A major version of zero is for the initial development.

- **Minor version**: A numeric identifier that gets increased if new features are added but the version is backward-compatible with the previous version and can be updated without breaking anything if you need the new functionality.

- **Patch**: A numeric identifier that gets increased if you release backward-compatible bug fixes. New patches should always be installed.

- **Pre-version**: A text identifier that is appended using a hyphen. The identifier must only use ASCII alphanumeric characters and hyphens (`[0-9A-Za-z-]`). The longer the text, the smaller the pre-version (meaning `-alpha` < `-beta` < `-rc`). A prerelease version is always smaller than a normal version (`1.0.0-alpha` < `1.0.0`).

It is considered best practice to prefix semantic versions in GitHub releases with `v` (for example, `v1.0`, `v1.0.1`, and so on).

See `https://semver.org/` for the complete specification regarding semantic versions.

There's more...

Versioning with tags opposes some kind of risk as everybody with write permissions to a repository can modify a tag. That's why, as the maintainer of a GitHub action, you are encouraged to use tag protection rules in addition to your branch protection rules (see `https://docs.github.com/en/repositories/managing-your-repositorys-settings-and-features/managing-repository-settings/configuring-tag-protection-rules`). A tag protection rule for `v*` will prevent everybody without admin permissions to your repository to modify tags starting with a `v`.

If you want to automate the process of creating semantic versions for your releases and the automatic creation of good release notes, then you can use **conventional commits** (see `https://www.conventionalcommits.org`). Conventional commits add a prefix to every commit, indicating if it is a feature or a fix and if it is breaking or not. You can combine this with **GitVersion** (see `https://gitversion.net/docs/`) to automatically create semantic versions for your release. You will learn more about this in *Chapter 7, Release Your Software with GitHub Actions*.

4

The Workflow Runtime

In this chapter, you will learn about the different runtime options for GitHub Actions. You will learn how to use different GitHub-hosted runners and how to set up self-hosted runners.

This chapter will cover the following topics:

- Setting up a self-hosted runner
- Auto-scaling self-hosted runners
- Scaling self-hosted runners with Kubernetes using **Actions Runner Controller (ARC)**
- Runners and runner groups
- GitHub-hosted runners
- Setting up a large runner
- Managing and auto-scaling ephemeral runners
- Security for GitHub-hosted and self-hosted runners

Technical requirements

For this chapter, you will need Docker and Visual Studio Code, though alternatively, you can use GitHub Codespaces. For the *Scaling self-hosted runners with Kubernetes using ARC* recipe, you will need either a Kubernetes cluster or an Azure subscription with the Azure CLI to set one up. For the recipes about runner groups, you will need a paid **Team** or **Enterprise** plan for a GitHub organization.

Setting up a self-hosted runner

So far, we have only used the `ubuntu-latest` label for our jobs. This runs the workflows on the latest version of a Ubuntu image hosted by GitHub. But there are also runners on macOS and Windows, with different configurations. You can host your own runners on any platform you like. In this first recipe, we will set up a self-hosted runner in a Linux Docker container. This way, it will be easy to scale it up and clean up the resources after our workflow run.

Getting ready...

You will need Docker installed for this recipe. You will also need to know your processor architecture. If you don't know it, just run `docker info` and look for `Architecture:`

```
$ docker info | grep Architecture
```

How to do it...

1. Go to a repository on GitHub. You can create a new one or you can use the `GitHubActionsCookbook` repository, which you created in *Chapter 1*. Go to **Settings | Actions | Runners** (`/settings/actions/runners`) and click on **New self-hosted runner** (see *Figure 4.1*):

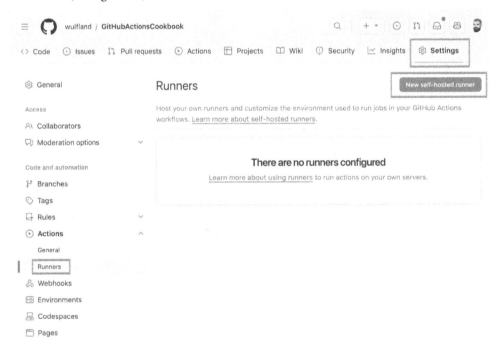

Figure 4.1 – Self-hosted runners can be added by going to a repository's Settings area

This will redirect you to `/settings/actions/runners/new`. Pick **Linux** for **Runner image** and set up the **Architecture** property according to your Docker environment.

Note the different scripts for each platform and processor architecture. You can copy the entire script for an installation on a VM. However, since we are in a Docker container, we will need some additional steps. Note that you can also copy individual lines (see *Figure 4.2*):

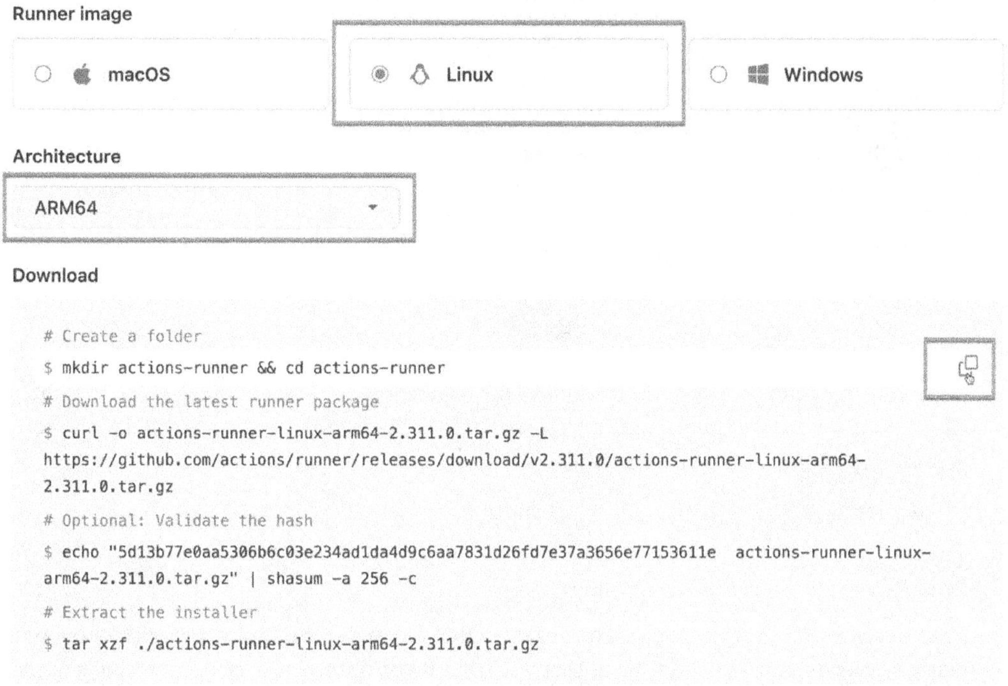

Runner image

○ 🍎 macOS	◉ 🐧 Linux	○ ⊞ Windows

Architecture

```
ARM64                                    ▼
```

Download

```
# Create a folder
$ mkdir actions-runner && cd actions-runner
# Download the latest runner package
$ curl -o actions-runner-linux-arm64-2.311.0.tar.gz -L
https://github.com/actions/runner/releases/download/v2.311.0/actions-runner-linux-arm64-
2.311.0.tar.gz
# Optional: Validate the hash
$ echo "5d13b77e0aa5306b6c03e234ad1da4d9c6aa7831d26fd7e37a3656e77153611e  actions-runner-linux-
arm64-2.311.0.tar.gz" | shasum -a 256 -c
# Extract the installer
$ tar xzf ./actions-runner-linux-arm64-2.311.0.tar.gz
```

Figure 4.2 – Script to install self-hosted runners on different platforms

2. Start a console in the latest version of an Ubuntu container:

```
$ docker run -it ubuntu:latest /bin/bash
```

3. Run the first line of the script. This will create a folder for the runner and change directory into it:

```
$ mkdir actions-runner && cd actions-runner
```

4. To download the runner binaries, we have to install `curl` in the container as this is not part of the normal Ubuntu image:

```
$ apt-get -y update; apt-get -y install curl
```

5. Now, execute the line from the script that downloads the latest runner package. Just copy and paste the line from your browser to your console:

```
$ curl -o actions-runner-{version}.tar.gz -L https://{URL}.tar.gz
```

6. Unzip the package using the `tar` command from the script:

```
tar xzf ./actions-runner-{version}.tar.gz
```

7. Install the dependencies that are needed for the runner by executing the following script:

```
$ ./bin/installdependencies.sh
```

8. Before we can configure the runner, we have to allow it to run as `root` as our container runs as root per default. We can do this by setting the RUNNER_ALLOW_RUNASROOT environment variable to a non-zero value:

```
$ export RUNNER_ALLOW_RUNASROOT="1"
```

9. Now, we can run the configuration script. If the other steps took too long for you to execute, then you might have to refresh the page in your browser as the tokens are only valid for a short period. Copy and execute the line containing the token:

```
./config.sh --url https://github.com/{OWNER}/{REPO} --token {TOKEN}
```

Press *Enter* and accept all default values.

After executing the script, you can navigate back to **Settings | Actions | Runners** to see the newly registered runner. You will see that it is still offline since we haven't started the runner process yet (see *Figure 4.3*):

Figure 4.3 – A configured runner that isn't running is shown as Offline

10. Start the runner using the following script:

```
$ ./run.sh
```

The runner will change to being in the `Idle` state, which means it's waiting for workflows to execute.

11. Create a simple workflow that uses the `self-hosted` label. The bash should be available on all platforms, so we can omit an additional label for `Linux`. However, you could also do this by running `[self-hosted, Linux]`:

```
name: Self-Hosted

on: [workflow_dispatch]

jobs:
  main:
    runs-on: self-hosted
    steps:
      - name: Output environment
        shell: bash
        run: |-
          echo "Runner Name: '${{ runner.name }}'"
          echo "Runner OS: '${{ runner.os }}'"
          echo "Runner ARCH: '${{ runner.arch }}'"
```

12. Execute the workflow and monitor your Docker container to see how it executes the workflow. You can repeat this step as many times as you want. So long as your container is running, it will execute all workflows with matching labels.

13. If you kill your container now, the runner will remain offline in GitHub. To remove it, navigate back to **Settings** | **Actions** | **Runners** and select **Remove runner** from the menu on the right-hand side of the runner (see *Figure 4.4*):

Figure 4.4 – Removing a runner from GitHub

Run the script provided by the dialogue to remove the runner:

```
$ ./config.sh remove --token {TOKEN}
```

The runner will now be gone from GitHub.

14. We can also configure the runner so that it only runs one job and then unregisters itself. This makes a lot of sense with containers. For this, you can just add the --ephemeral switch to the configuration step after generating a new token:

```
./config.sh --url {URL} --token {TOKEN} --ephemeral
```

15. Run the workflow again; you will see that the runner will be removed after execution.

16. Next, we can put everything we've learned into a Dockerfile (you can use the file available at https://github.com/wulfland/GitHubActionsCookbook/blob/main/SelfHostedRunner/Dockerfile). This way, we can create a reusable Docker image that will register itself, wait for a job, execute it, and then terminate every time you run it.

We inherit from ubuntu:latest for simplicity. You can easily replace this with a base image that contains all your build tools:

```
FROM ubuntu:latest
```

Set the variables that will be used for the connection. Leave TOKEN and RUNNER_NAME empty as these values will be provided when the container starts, not during image creation. Set the correct URL, platform, and version:

```
ENV TOKEN=
ENV RUNNER_NAME=
ENV RUNNER_URL="https://github.com/{owner}/{repo}"
ENV GH_RUNNER_PLATFORM="linux-arm64"
ENV GH_RUNNER_VERSION="2.311.0"
ENV LABELS="self-hosted,ARM64,Linux"
ENV RUNNER_GROUP="Default"
```

Before installing the missing packages, we have to set DEBIAN_FRONTEND to noninteractive to ensure that the operating system does not prompt for user input during the Docker image build process:

```
ARG DEBIAN_FRONTEND=noninteractive
```

To have fewer layers in the Docker image, it is best to combine the entire script into one RUN command. The script updates the package manager and all its packages and installs all dependencies, adds the docker user that the container will run under (we don't want the container to run as root), downloads the corresponding package, unzips it, changes the owner to the docker user, and executes the installdependencies.sh script:

```
RUN apt-get -y update && \
    apt-get upgrade -y && \
    useradd -m docker && \
    apt-get install -y --no-install-recommends curl
ca-certificates && \
```

```
mkdir -p /opt/hostedtoolcache /home/docker/actions-runner && \
curl -L https://github.com/actions/runner/releases/download/
v${GH_RUNNER_VERSION}/actions-runner-${GH_RUNNER_PLATFORM}-${GH_
RUNNER_VERSION}.tar.gz -o /home/docker/actions-runner/actions-
runner.tar.gz && \
tar xzf /home/docker/actions-runner/actions-runner.tar.gz -C /
home/docker/actions-runner && \
chown -R docker /home/docker && \
/home/docker/actions-runner/bin/installdependencies.sh
```

Run the container as the docker user and set the working directory to actions-runner in the home directory of that user:

```
USER docker
WORKDIR /home/docker/actions-runner
```

If the container will run, we must check if TOKEN and RUNNER_NAME are provided. Then, we must run the config script with all parameters – including --ephemeral – and then run the run.sh script:

```
CMD if [ -z "$TOKEN" ]; then echo 'TOKEN is not set'; exit 1; fi
&& \
    if [ -z "$RUNNER_NAME" ]; then echo 'RUNNER_NAME is not set';
exit 1; fi && \
    ./config.sh --url "${RUNNER_URL}" --token "${TOKEN}"
--name "${RUNNER_NAME}" --work "_work" --labels "${LABELS}"
--runnergroup "${RUNNER_GROUP}" --unattended --ephemeral && \
    ./run.sh
```

17. Step into the folder containing the Dockerfile and create a Docker image from the file:

```
$ docker build -t simple-ubuntu-runner .
```

18. Now, run as many instances of the image as you like with docker run. The -d (--detached) option will run the container in detached mode in the background. It will not block your console, but it will not receive input or display output in the terminal. The --rm option will remove the container when it exits. Pass in arguments for RUNNER_NAME and TOKEN using the -e option. Keep in mind that these names are case-sensitive!

```
$ docker run -d --rm -e RUNNER_NAME=Runner1 -e TOKEN={TOKEN}
simple-ubuntu-runner
```

You will see the runners in the **Settings** area of your repository, as shown in *Figure 4.5*:

Figure 4.5 – Running the same Docker image multiple times to
get ephemeral runners waiting for incoming jobs

Start the workflow as many times as you have created containers to run a workflow and clean up the container after execution.

How it works...

Let's understand how the flow works.

The self-hosted runner application

A self-hosted runner is created by installing the open source runner application (https://github.com/actions/runner). The application is based on the .NET Core runtime and can run on a large number of operating systems and processor architectures. It can run on *macOS 11* (Big Sur) or later, *Windows* (7 to 10 and Server 2012 R2 to 20222), and many Linux distributions (Red Hat Enterprise 7 or later, Fedora 29 or later, Ubuntu 16.04 or later, and many more). It can also run on x64, ARM64, and ARM32. For an up-to-date list of supported operating systems, see https://docs.github.com/en/actions/hosting-your-own-runners/managing-self-hosted-runners/about-self-hosted-runners#supported-architectures-and-operating-systems-for-self-hosted-runners. You can use the bin/installdependencies.sh script in the runner applications folder to install all the required libraries for the .NET Core runtime.

If you want to run Docker-based actions, you must use a Linux image. Windows and macOS are not supported for running Docker-based actions!

Authentication to GitHub

Connecting the runner to GitHub is done using a **configuration token** that can be generated by a user through the GitHub UI. The token is only valid for 1 hour and you can only use the token to install runners. You can also create an installation token through the REST API on demand by sending a POST request to `https://api.github.com/repos/{OWNER}/{REPO}/actions/runners/registration-token` (or `https://api.github.com/orgs/{ORG}/actions/runners/registration-token` for runners at the organization level).

Here's an example of how you would receive a token using a **personal access token** (**PAT**):

```
$ curl -L \
> -X POST \
> -H "Accept: application/vnd.github+json" \
> -H "Authorization: Bearer <YOUR-PAT>" \
> -H "X-GitHub-Api-Version: 2022-11-28" \
> https://api.github.com/repos/{OWNER}/{REPO}/actions/runners/
registration-token
```

The result also contains the expiration date. If you want to use the token in a variable, pipe the result to `jq`, like this:

```
$ TOKEN=$(<curl command> | jq .token --raw-output)
```

You must authenticate using a PAT access token with the `repo` scope to use this endpoint. GitHub Apps must have administration permission for repositories and the `organization_self_hosted_runners` permission for organizations. Authenticated users must have admin access to repositories or organizations, or the `manage_runners:enterprise` scope for enterprises.

The token is only valid for registration. During the registration process, a **JWT** (**JSON Web Token for OAuth exchange**) will be received from the server that only has permission to listen to the queue. When a workflow run starts, another pre-built token with a limited scope (defined by the workflow) will be created for the life of the build. That token can't be accessed via ad hoc scripts or untrusted code – only by the build agent and tasks. The RSA private key for the OAuth token exchange between the agent and server will be stored in a file called `.credentials_rsaparams` and the server holds the public key. Every 50 minutes, the server will send a new token to the agent that's encrypted by the public key. The OAuth configuration is stored in the `.credentials` file:

```
{
  "scheme": "OAuth",
  "data": {
    "clientId": "{CLIENT_ID}",
    "authorizationUrl": "https://pipelinesghubeus4.actions.
githubusercontent.com/{TOKEN}/_apis/oauth2/token",
    "requireFipsCryptography": "True"
  }
```

Running the application as a service

On Windows, the configuration script will ask you if you want to execute the runner as a service so that it will start alongside the environment. On Linux, you will have to configure the service yourself using the `svc.sh` script:

```
sudo ./svc.sh install
sudo ./svc.sh start
```

Network communication

The runner application communicates with GitHub using an outgoing HTTPS connection on **port 443** using **long polling** with a 50-second timeout. This means that the application asks GitHub if any work is queued for the labels of the runner and then waits for 50 seconds for a response before the connection is closed. Immediately after closing the connection, a new connection is started. There is no need for any inbound connection from GitHub or to open any firewall ports. Only secure outbound connections using SSL over port 443.

Updating self-hosted runners

Self-hosted runners will automatically check if there is a new version of the runner application available and update it. GitHub will only update the runner itself – the rest of the machine is managed by the customer.

Cleaning up

It is worth noting that the GitHub runner application will not clean up resources after a workflow run. This behavior is different from GitHub-hosted runners as they provide an ephemeral fresh environment for every workflow run. If you download your repository and perform a build, all files will just stay there. If you want to use a runner application for multiple workflow runs, then you have to clean up everything yourself. That's why I emphasize the use of ephemeral runners in containers. This way, you always have a clean environment.

You can use workflow logic to clean up after your workflow runs – but you can also use pre- or post-job scripts to do this on the runner. To configure a pre- or post-job script, you need to save a script file in a location to which the runner has access, and then configure an environment variable with one of the following names and the full path to the script as the value:

- `ACTIONS_RUNNER_HOOK_JOB_STARTED`
- `ACTIONS_RUNNER_HOOK_JOB_COMPLETED`

As an alternative, you can store key-value pairs in a `.env` file inside the runner application directory.

There's more...

Installing the runner on macOS is the same as it is for Linux. The difference on Windows is that the script is a PowerShell script instead of a bash script. It uses `Invoke-WebRequest` instead of `curl`, for example, but all the steps are the same. The scripts to configure and start the runner have the `.cmd` extension instead of `.sh`:

```
$ ./config.cmd --url <URL> --token <TOKEN>
$ ./run.cmd
```

If you've successfully installed the runner in a Linux container, then you will have no problem installing it on Windows.

Auto-scaling self-hosted runners

In this recipe, we'll be building on the previous recipe so that we have a solution that automatically starts a new instance of the ephemeral Docker container every time a new workflow is triggered. We'll use a GitHub webhook for that.

Getting ready...

Make sure you still have the `simple-ubuntu-runner` Docker image we created in the previous recipe on your machine or GitHub Codespaces.

How to do it...

1. Go to `https://github.com/settings/apps` and click on **New GitHub App**.

2. Set **GitHub App Name** to `auto-scale-runners` and **Homepage URL** to the URL of the repository you are using (see *Figure 4.6*):

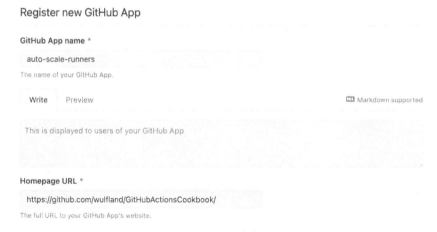

Figure 4.6 – Setting the name and URL for the new app

3. Skip the **Identifying and authorizing users** and **Post installation** section and proceed to **Webhook**.

4. Open another browser tab, go to https://smee.io, and click on **Start new channel**. Copy the **Webhook Proxy URL** value.

5. Go back to the other tab and paste the URL into the **Webhook URL** field. Set **Webhook secret** to a string that you will remember later (see *Figure 4.7*):

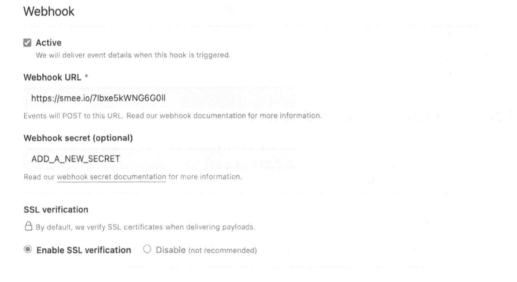

Figure 4.7 – Configuring the webhook

6. Under **Permissions | Repository permissions**, set **Actions** to **Read-only** and **Administration** to **Read and write** (see *Figure 4.8*):

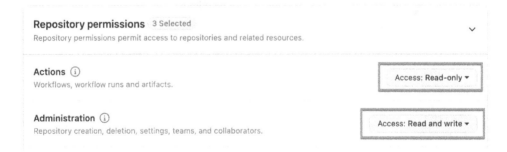

Figure 4.8 – Configuring Repository permissions for the app

7. Under **Subscribe to events**, select **Workflow job** (see *Figure 4.9*):

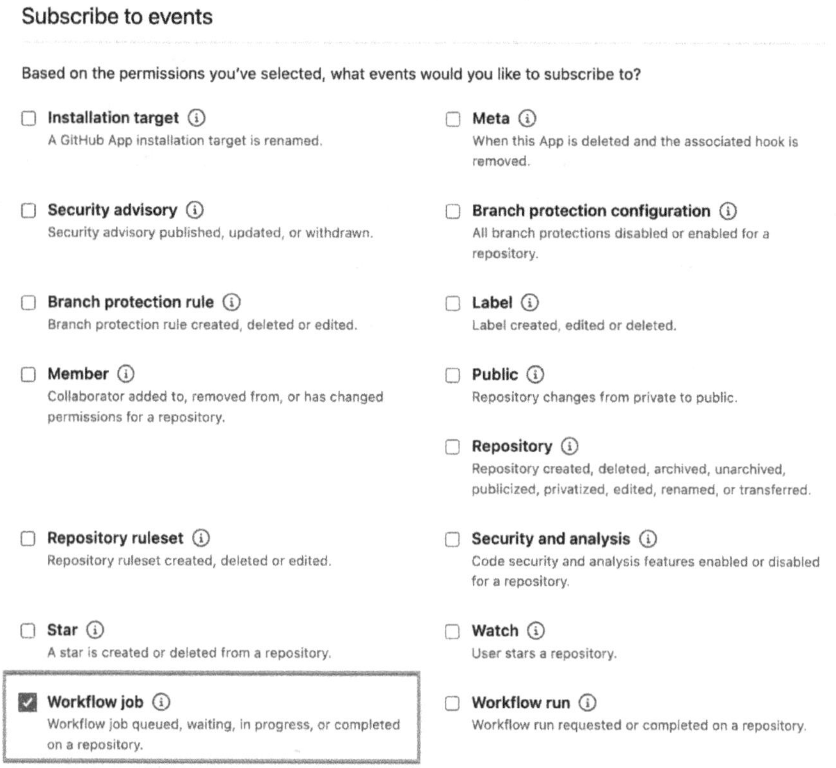

Subscribe to events

Based on the permissions you've selected, what events would you like to subscribe to?

☐ **Installation target** ⓘ
A GitHub App installation target is renamed.

☐ **Meta** ⓘ
When this App is deleted and the associated hook is removed.

☐ **Security advisory** ⓘ
Security advisory published, updated, or withdrawn.

☐ **Branch protection configuration** ⓘ
All branch protections disabled or enabled for a repository.

☐ **Branch protection rule** ⓘ
Branch protection rule created, deleted or edited.

☐ **Label** ⓘ
Label created, edited or deleted.

☐ **Member** ⓘ
Collaborator added to, removed from, or has changed permissions for a repository.

☐ **Public** ⓘ
Repository changes from private to public.

☐ **Repository** ⓘ
Repository created, deleted, archived, unarchived, publicized, privatized, edited, renamed, or transferred.

☐ **Repository ruleset** ⓘ
Repository ruleset created, deleted or edited.

☐ **Security and analysis** ⓘ
Code security and analysis features enabled or disabled for a repository.

☐ **Star** ⓘ
A star is created or deleted from a repository.

☐ **Watch** ⓘ
User stars a repository.

☑ **Workflow job** ⓘ
Workflow job queued, waiting, in progress, or completed on a repository.

☐ **Workflow run** ⓘ
Workflow run requested or completed on a repository.

Figure 4.9 – Subscribing to the workflow job webhook

8. Under **Where can this GitHub App be installed**, select **Only on this account** and click **Create GitHub App**.

9. In your newly created app, click on **Generate a private key**. The private key will be automatically downloaded. Move it to your repository.

10. Copy the **App ID** value from the **General** tab of the app (see *Figure 4.10*):

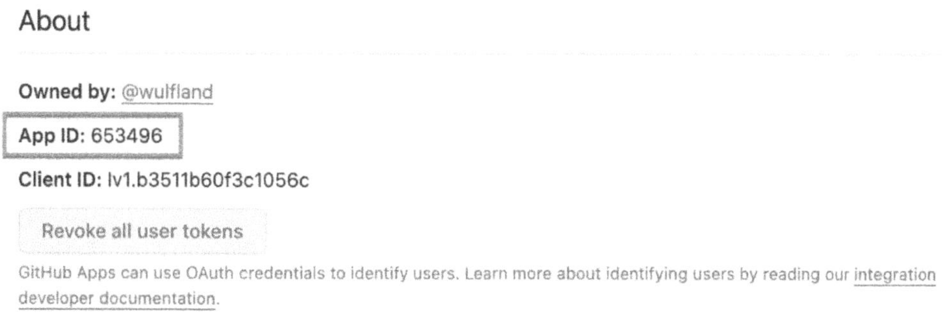

About

Owned by: @wulfland

App ID: 653496

Client ID: Iv1.b3511b60f3c1056c

Revoke all user tokens

GitHub Apps can use OAuth credentials to identify users. Learn more about identifying users by reading our integration developer documentation.

Figure 4.10 – Getting the App ID value

11. In your repository, create a new file called `.env` and add variables for `APP_ID`, `WEBHOOK_SECRET`, and `PRIVATE_KEY_PATH` with the corresponding values:

```
APP_ID="653496"
WEBHOOK_SECRET="YOUR_SECRET"
PRIVATE_KEY_PATH="auto-scale-runners.2023-11-26.private-key.pem"
```

We will use these environment variables in our application later to authenticate to GitHub.

12. Add the `.env` file to the `.gitignore` file so that you don't accidentally commit this:

```
$ echo ".env" >> .gitignore
```

13. In the app in GitHub, select **Install App** and click **Install** (see *Figure 4.11*):

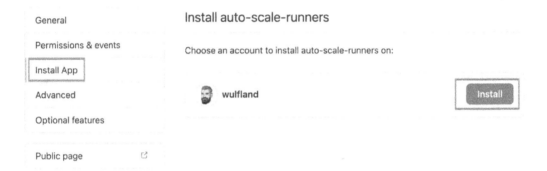

Figure 4.11 – Installing the app

14. In the dialogue that appears, select your repository and click **Install** (see *Figure 4.12*):

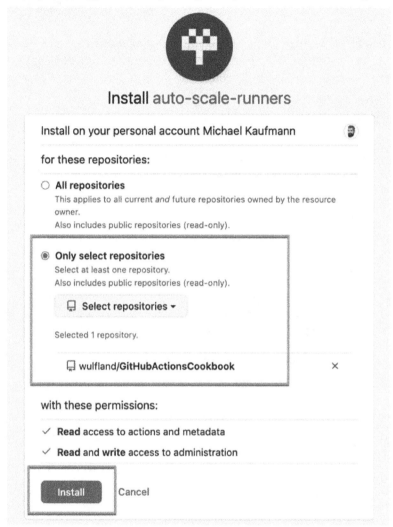

Figure 4.12 – Installing the app in your repository

15. Next, we will create the server that will run when it receives a payload from the webhook. Go to your repository and initialize it:

```
$ npm init -yes
```

Add the dependencies we are using in the code:

```
$ npm install octokit
$ npm install dotenv
$ npm install smee-client --save-dev
```

Add the node_modules folder to .gitignore:

```
$ echo "node_modules" >> .gitignore
```

16. Create a new file called app.js. You can copy the content from here: https://github.com/wulfland/GitHubActionsCookbook/blob/main/SelfHostedRunner/auto-scale/app.js. However, I will still go through the code here step by step.

17. Add the necessary dependencies:

```
import dotenv from "dotenv";
import {App} from "octokit";
import {createNodeMiddleware} from "@octokit/webhooks";
import fs from "fs";
import http from "http";
import { exec } from 'child_process';
```

18. Next, use the dotenv package to read the environment variables from the .env file we created earlier:

```
dotenv.config();
const appId = process.env.APP_ID;
const webhookSecret = process.env.WEBHOOK_SECRET;
const privateKeyPath = process.env.PRIVATE_KEY_PATH;
```

19. Next, load the private key from the corresponding path:

```
const privateKey = fs.readFileSync(privateKeyPath, "utf8");
```

20. Create a new instance of the app class from octokit:

```
const app = new App({
  appId: appId,
  privateKey: privateKey,
  webhooks: {
    secret: webhookSecret
  },
});
```

Then, register an event handler for the workflow_job.queued event:

```
app.webhooks.on("workflow_job.queued",
handleNewQueuedJobsRequestOpened);
```

21. Call the GitHub API to receive a new token that can register runners. We need the token so that we can pass it to our container. We must use workflow_job.id to create a unique name for our runner:

```
const response = await octokit.request('POST /repos/{owner}/
{repo}/actions/runners/registration-token', {
  owner: payload.repository.owner.login,
```

```
        repo: payload.repository.name,
        headers: {
          'X-GitHub-Api-Version': '2022-11-28'
        }
    });


    const token = response.data.token;
    const runner_name = `Runner_${payload.workflow_job.id}`;
```

22. Then, we must create a new instance of our Docker container and pass in the token and name:

```
    exec(`docker run -d --rm -e RUNNER_NAME=${runner_name} -e
    TOKEN=${token} simple-ubuntu-runner`, (error, stdout, stderr) =>
    {
      if (error) {
        console.error(`exec error: ${error}`);
        return;
      }
      console.log(`stdout: ${stdout}`);
      console.error(`stderr: ${stderr}`);
    });
```

I've skipped the error handling part here as it isn't relevant.

23. In the last part of the file, we must create a development server that will listen on port 3000:

```
    const port = 3000;
    const host = 'localhost';
    const path = "/api/webhook";
    const localWebhookUrl = `http://${host}:${port}${path}`;
    const middleware = createNodeMiddleware(app.webhooks, {path});
    http.createServer(middleware).listen(port, () => {
      console.log(`Server is listening for events at:
    ${localWebhookUrl}`);
      console.log('Press Ctrl + C to quit.')
    });
```

24. In the package.json file, add a top-level entry called type and set it to module. Then, add a script called server that will run the application:

```
    "type": "module",
    "scripts": {
      "server": "node app.js"
    },
```

25. We are ready! Open a new terminal and start a smee client with the URL of your channel (*Step 4*):

```
    $ npx smee -u https://smee.io/{ID} -t http://localhost:3000/api/
    webhook
```

In a terminal in your repository, run the app:

```
$ npm run server
```

Start a new workflow run for the `Self-Hosted` workflow:

```
$ gh workflow run Self-Hosted
```

Note how your `smee` client receives the webhook that was forwarded from GitHub and how your server processes it and starts a new container that executes your workflow.

How it works...

GitHub Apps provide you with an easy way to authenticate to GitHub and register to webhooks. In this recipe, we registered to the `workflow_job` webhook with the `queued` action type so that every time a new workflow is queued, we can start a new runner. If you have larger images that take longer to load, you could also have a pool of existing runners and still load a new one if a job gets queued. The ephemeral runner you start does not have to be the one executing your job.

As we needed an endpoint that could be reached by GitHub, we used `smee.io` as a proxy and had it forward the payload in case of an event. This is not meant for production use. It just gives us a convenient way to develop locally or in a GitHub Codespace, without the need to have a publicly available inbound port. For production use, you should host the application on a web server.

There's more...

This recipe intended to provide you with the basic building blocks to create your own solution for scaling self-hosted runners. Using ephemeral runners and webhooks, it is easy to automate this process. But if you need a more scalable, mature solution, then you should probably look into doing this with Kubernetes.

Scaling self-hosted runners with Kubernetes using ARC

Kubernetes is very powerful but also quite complex. Be aware that in this recipe, I will only focus on getting you started when it comes to scaling self-hosted runners in Kubernetes. If you want to run and maintain a secure environment, you will need deeper Kubernetes know-how and must take on more work, depending on your needs.

ARC is a Kubernetes operator that orchestrates and scales your self-hosted runners' workloads. It is an open source project but it is now fully supported by GitHub.

Getting ready...

If you already have a Kubernetes cluster, you can use that. If not, you can create a new one in Azure by running the following commands:

```
$ az group create --name AKSCluster -l westeurope
$ az aks create --resource-group AKSCluster \
> --name AKSCluster \
> --node-count 3 \
> --enable-addons monitoring \
> --generate-ssh-keys
$ az aks get-credentials --resource-group AKSCluster --name AKSCluster
```

Make sure you have `cert-manager` installed in the cluster (`https://cert-manager.io/docs/installation/`). You can do this by running the following command. Make sure you replace the version number with an up-to-date version:

```
$ kubectl apply -f https://github.com/cert-manager/cert-manager/
releases/download/v1.13.2/cert-manager.yaml
```

You can check if the prerequisites have changed in the quick-start tutorial for ARC: `https://github.com/actions/actions-runner-controller/blob/master/docs/quickstart.md`.

How to do it...

1. Deploy ARC to your cluster. Make sure you change its version to an up-to-date version:

    ```
    $ kubectl apply -f https://github.com/actions/actions-runner-
    controller/releases/download/v0.23.7/actions-runner-controller.
    yaml
    ```

2. Create a PAT in GitHub with the `repo` scope. Go to `https://github.com/settings/tokens/new`, select **repo**, set the expiration date, and click **Generate token**. Then, copy the token.

3. Now, save the token as a secret in Kubernetes:

    ```
    $ kubectl create secret generic controller-manager \
    > -n actions-runner-system \
    > --from-literal=github_token=<YOUR_TOKEN>
    ```

4. Create a file called `runnerdeployment.yml` in your repository with the following content. Replace the repository owner and name with the values for your repository:

```
apiVersion: actions.summerwind.dev/v1alpha1
kind: RunnerDeployment
metadata:
  name: example-runnerdeploy
spec:
  replicas: 1
  template:
    spec:
      repository: wulfland/GitHubActionsCookbook
```

5. Apply `RunnerDeployment` to your cluster:

```
$ kubectl apply -f runnerdeployment.yml
```

6. Now, you should have one runner and two pods running:

```
$ kubectl get runners
$ kubectl get pods
```

Verify that you can see the runner in GitHub (`settings/actions/runners`). Note that the name will change after every workflow run (see *Figure 4.13*):

Runners

New self-hosted runner

Host your own runners and customize the environment used to run jobs in your GitHub Actions workflows. Learn more about self-hosted runners.

Runners				Status
⊞ **example-runnerdeploy-j92xb-q8mwb**	self-hosted	Linux	X64	● Idle ⋯

Figure 4.13 – The ARC runner in GitHub

7. Run your `.github/workflows/self-hosted.yml` workflow from the previous recipes and inspect the output. Note that it is executed in your AKS cluster (see *Figure 4.14*):

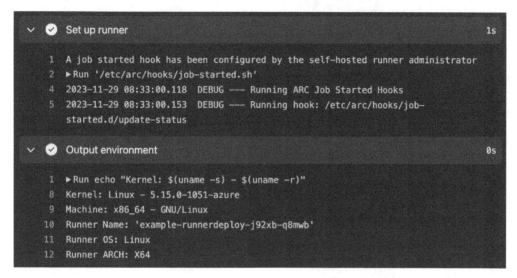

```
  ⌄  ✔  Set up runner                                                          1s

    1  A job started hook has been configured by the self-hosted runner administrator
    2  ▶ Run '/etc/arc/hooks/job-started.sh'
    4  2023-11-29 08:33:00.118  DEBUG ---- Running ARC Job Started Hooks
    5  2023-11-29 08:33:00.153  DEBUG ---- Running hook: /etc/arc/hooks/job-
       started.d/update-status

  ⌄  ✔  Output environment                                                     0s

    1  ▶ Run echo "Kernel: $(uname -s) - $(uname -r)"
    8  Kernel: Linux - 5.15.0-1051-azure
    9  Machine: x86_64 - GNU/Linux
   10  Runner Name: 'example-runnerdeploy-j92xb-q8mwb'
   11  Runner OS: Linux
   12  Runner ARCH: X64
```

Figure 4.14 – The self-hosted workflow is executed by ARC in the Kubernetes cluster

How it works...

ARC runners are set up as ephemeral by default. They use Kubernetes replica sets and spin up a new container after execution automatically. ARC provides three options to scale:

- **Scheduled**: Scale up and down based on a schedule
- **Scale based**: The percentage of runners that are busy executing a job
- **On-demand**: Start an instance when a new workflow job is queued

ARC supports creating runners at the enterprise, organization, and repository levels. You can authenticate at the organization and repository levels using a GitHub app or PAT. However, at the enterprise level, you *have* to use a PAT as apps cannot be scoped to an enterprise.

You can configure scale sets with different images and namespaces for different teams. This also allows you to limit networking access between them.

There's more...

Attacks on the software supply chain and the build process oppose a big thread. You should be very careful, especially with self-hosted runners. Never use them with public repositories that allow forking and take measures to control all the dependencies you pull on your runners so that nobody can do the following:

- Tamper with files in your build process
- Escape your container/sandbox or access files from outside the workflow scope

- Compromise your dependencies by manipulating the dependency cache (cache poisoning)

- Exfiltrate data or secrets and send data to a control server

You can use the **Harden-Runner** action from StepSecurity (`https://github.com/step-security/harden-runner`) in your workflows. It will monitor your workflow on policies such as outbound network egress:

```
steps:
  - uses: step-security/harden-runner@v2.6.1
    with:
      egress-policy: audit
```

You can also use it to block traffic and only allow certain endpoints and ports:

```
egress-policy: block
allowed-endpoints: >
  api.nuget.org:443
  github.com:443
```

For ARC, instead of adding the Harden-Runner action to each workflow job, you can install the ARC Harden-Runner DaemonSet on your Kubernetes cluster. The DaemonSet will constantly monitor each workflow run without the need to add the action to each workflow.

You can access security insights and runtime detections under the **Runtime Security** tab in your dashboard.

Please note that this is not free software. There is a free community license for public repositories on `https://github.com`, but for private repositories or **GitHub Enterprise Server** (**GHES**), you will have to purchase a license (see `https://www.stepsecurity.io/pricing`).

Runners and runner groups

At the organization and enterprise levels, access to runners is organized in runner groups. The association from workflow to runner is done by labels – but runner groups control what runners a workflow has access to.

Getting ready...

Please note that in free organizations, there is only one runner group, called Default, that you can use to add self-hosted runners. To create multiple runner groups or use them for GitHub-hosted runners, you will need a paid Team or Enterprise plan.

How to do it...

1. In an organization with a paid plan, navigate to **Settings | Actions | Runner groups** (`/settings/actions/runner-groups`) and click **New group**.

    Give the group a name. Under **Repository access**, change the selection from **All repositories** to **Selected repositories** and click the gear icon to select one or multiple repositories that will have access to the group (see *Figure 4.15*). Note that you can allow access to public repositories here but that this option is disabled by default:

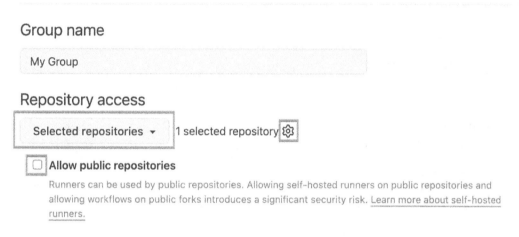

Figure 4.15 – Managing access to a group for repositories

2. Under **Workflow access**, pick **Selected workflows** and click the gear icon (see *Figure 4.16*):

Workflow access

Control how these runners are used by restricting them to specific workflows. Learn more about managing runner groups.

Selected workflows ▾   0 selected workflows ⚙

Figure 4.16 – Limiting access to certain workflows

Note that in the dialogue that appears, you can add multiple patterns to identify workflows (see *Figure 4.17*):

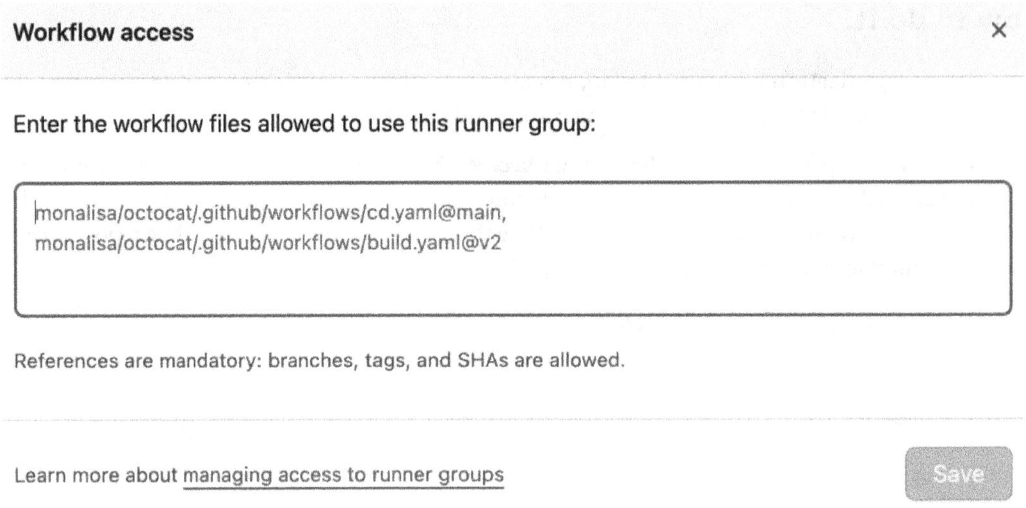

Workflow access   ✕

Enter the workflow files allowed to use this runner group:

monalisa/octocat/.github/workflows/cd.yaml@main,
monalisa/octocat/.github/workflows/build.yaml@v2

References are mandatory: branches, tags, and SHAs are allowed.

Learn more about managing access to runner groups   Save

Figure 4.17 – Syntax for limiting access to workflow versions

Workflows are specified with the path to the workflow file and a valid git reference. The reference can be a branch, tag, or SHA value.

3. Exit the dialogue, set the value back to **All workflows**, and click **Create group**. You can add now self-hosted runners and GitHub-hosted runners to the group. We'll cover GitHub-hosted runners in the next recipe. Click **New self-hosted runner** (see *Figure 4.18*):

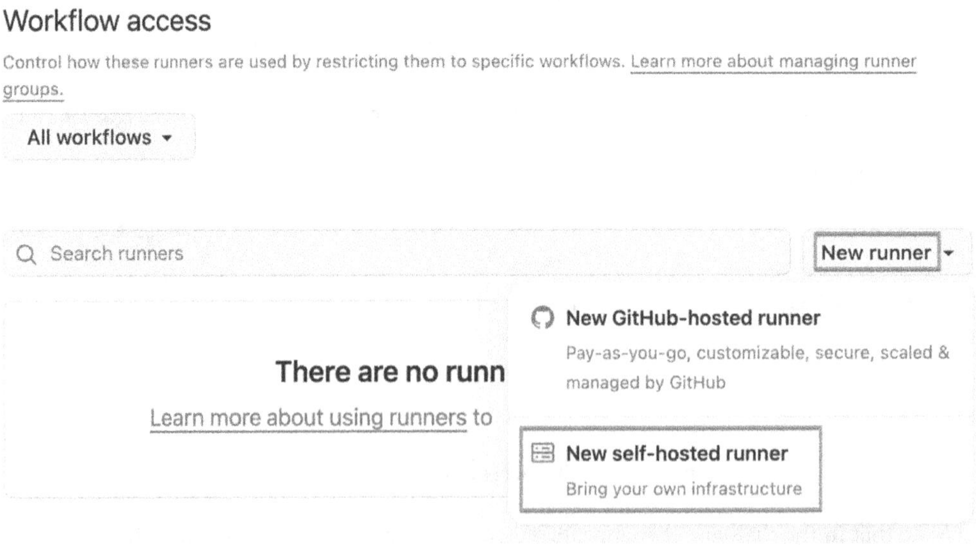

Workflow access

Control how these runners are used by restricting them to specific workflows. Learn more about managing runner groups.

All workflows ▾

Q Search runners   New runner ▾

                                                        ◯ New GitHub-hosted runner
                                                            Pay-as-you-go, customizable, secure, scaled &
                      There are no runn   managed by GitHub

            Learn more about using runners to
                                                        ▦ New self-hosted runner
                                                            Bring your own infrastructure

Figure 4.18 – Adding self-hosted runners to a runner group

4. Note that the dialogue for adding runners is the same as what appeared in the *Setting up a self-hosted runner* recipe. You can test this with your container by overriding RUNNER_URL:

```
$ docker run -d --rm -e RUNNER_NAME=Runner_Group \
> -e TOKEN={TOKEN} \
> -e RUNNER_URL=https://github.com/{org} \
> simple-ubuntu-runner
```

This will create the runner in the Default group! Open the runner and assign it to the new group you created (see *Figure 4.19*):

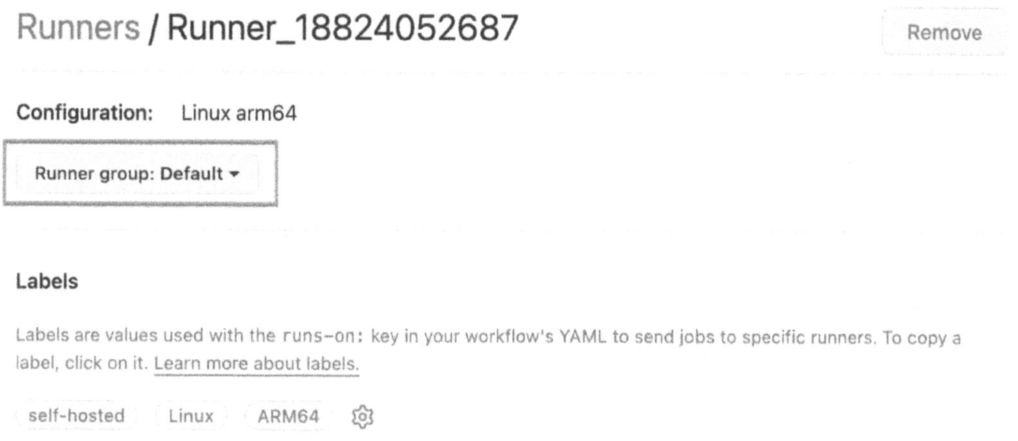

Runners / Runner_18824052687   Remove

Configuration: Linux arm64

Runner group: Default ▾

Labels

Labels are values used with the runs-on: key in your workflow's YAML to send jobs to specific runners. To copy a label, click on it. Learn more about labels.

self-hosted   Linux   ARM64   ⚙

Figure 4.19 – Assigning a runner to a runner group

Note that a runner can only be assigned to one group. To directly assign the runner to a group while setting it up, you have to pass the parameter to the config script, which is not in the instructions:

```
$ ./config.sh --url $org_or_enterprise_url --token $token
--runnergroup rg-runnergroup
```

Runner groups are an important feature for organizations and enterprises to manage access to runners. The feature is straightforward and does not need a lot of explaining – that's why I'd rather keep it short and simple. In the next recipe, we will use runner groups to add larger GitHub-hosted runners.

GitHub-hosted runners

In this last recipe of this chapter, we'll be creating a larger GitHub-hosted runner with network isolation in the runner group.

Getting ready...

You need the runner group you created in the previous recipe in an enterprise or organization with a paid plan!

How to do it...

1. In the runner group you created, click **New runner | New GitHub-hosted runner** (see *Figure 4.20*):

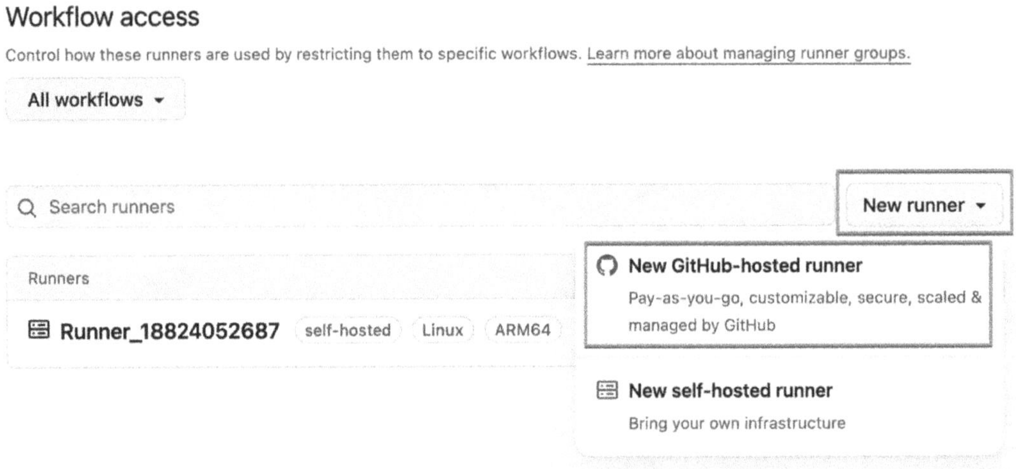

Workflow access

Control how these runners are used by restricting them to specific workflows. Learn more about managing runner groups.

All workflows ▾

Q Search runners   New runner ▾

Runners   ○ **New GitHub-hosted runner**
                                                            Pay-as-you-go, customizable, secure, scaled &
▤ **Runner_18824052687**   self-hosted   Linux   ARM64   managed by GitHub

                                                         ▤ **New self-hosted runner**
                                                            Bring your own infrastructure

Figure 4.20 – Creating new GitHub-hosted runners

2. Give the runner a name. The name must be between 1 and 100 characters, and it may only contain uppercase letters (A-Z) and lowercase letters (a-z), numbers (0-9), dots (.), dashes (-), and underscore (_). Pick a **Runner image** value (**Ubuntu** or **Windows**) and the corresponding version (see *Figure 4.21*):

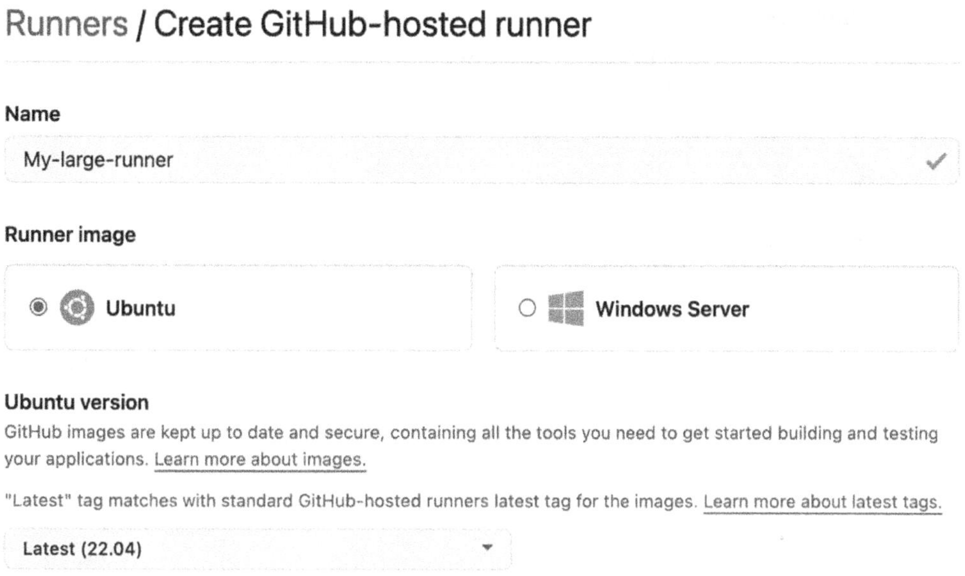

Runners / Create GitHub-hosted runner

Name

My-large-runner   ✓

Runner image

◉ Ubuntu   ○ Windows Server

Ubuntu version

GitHub images are kept up to date and secure, containing all the tools you need to get started building and testing your applications. Learn more about images.

"Latest" tag matches with standard GitHub-hosted runners latest tag for the images. Learn more about latest tags.

Latest (22.04)   ▼

Figure 4.21 – Configuring the name and image for the runner

3. Pick the size of the runner (see *Figure 4.22*). Note that larger runners are more expensive. See *Chapter 1* for details on the pricing of larger runners:

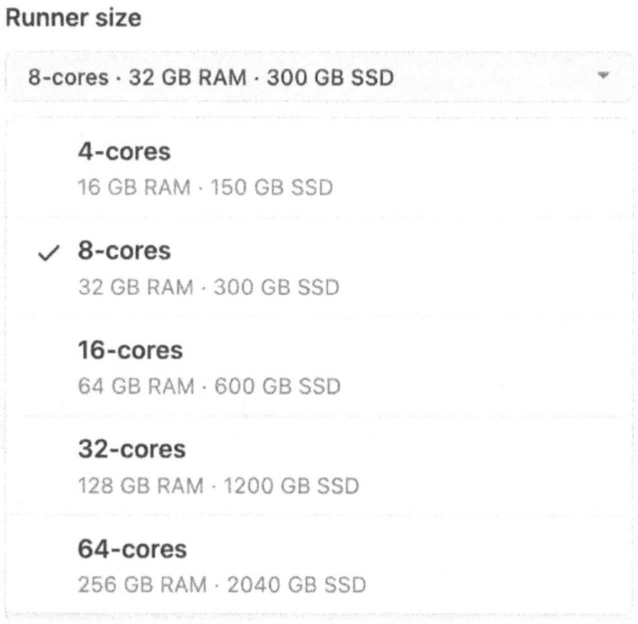

Runner size

8-cores · 32 GB RAM · 300 GB SSD   ▼

4-cores
16 GB RAM · 150 GB SSD

✓ **8-cores**
32 GB RAM · 300 GB SSD

16-cores
64 GB RAM · 600 GB SSD

32-cores
128 GB RAM · 1200 GB SSD

64-cores
256 GB RAM · 2040 GB SSD

Figure 4.22 – Picking the size for the new runner

4. You can limit the maximum number of concurrent jobs. The maximum is 500. Leave it set at the default of 50. Also, leave the runner group that was automatically set to the group in which you started creating the runner (see *Figure 4.23*):

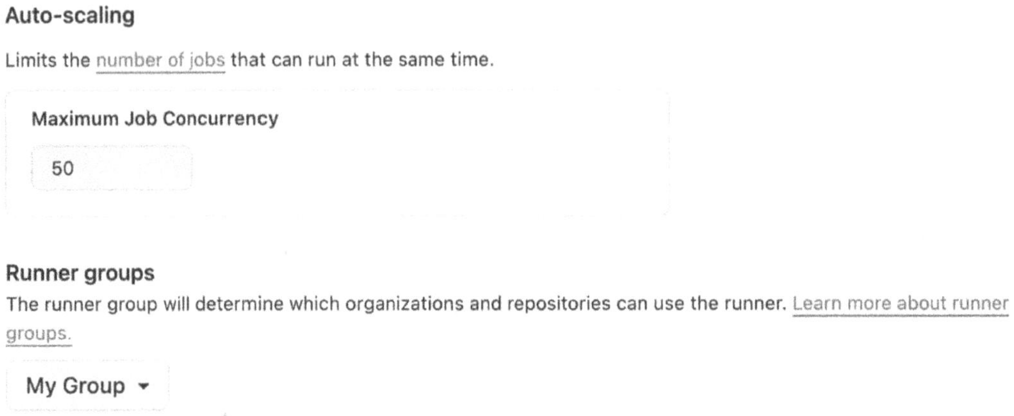
Auto-scaling
Limits the number of jobs that can run at the same time.
Maximum Job Concurrency
50
Runner groups
The runner group will determine which organizations and repositories can use the runner. Learn more about runner groups.
My Group

Figure 4.23 – Setting job concurrency and a runner group

5. You can enable network isolation by assigning a unique and static public IP address range to the runner (see *Figure 4.24*). Hit **Create runner** to finish the process:

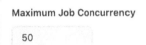

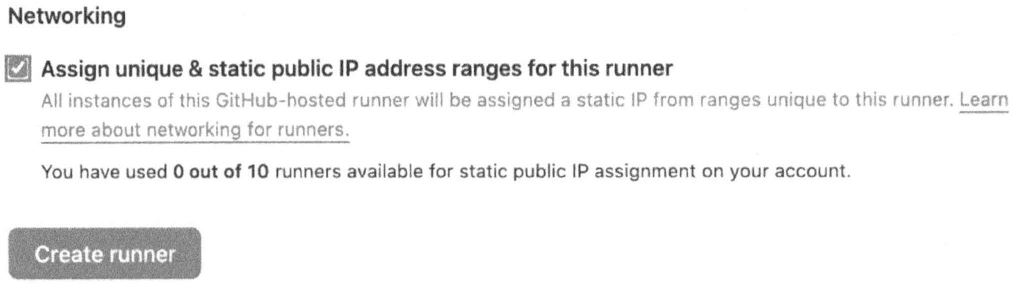
Networking
Assign unique & static public IP address ranges for this runner
All instances of this GitHub-hosted runner will be assigned a static IP from ranges unique to this runner. Learn more about networking for runners.
You have used 0 out of 10 runners available for static public IP assignment on your account.
Create runner

Figure 4.24 – Enabling network isolation for runners

6. Note that provisioning takes some time. Once the runner is ready, you can inspect the IP range associated with it, at which point you will see that the label is the same as the runner's name (see *Figure 4.25*). You can now start executing workflows on your larger runner!

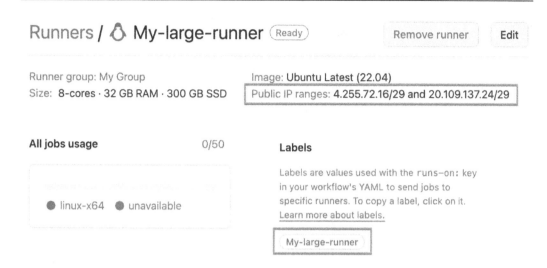

Figure 4.25 – Larger runners with network isolation

How it works...

GitHub will provision the larger runners for you and it will assign a static public IP range if you wish. Note that you have to use the runner; if not, GitHub will shut it down after some time. Network isolation allows you to give the runner access to local resources without the need for public access.

5

Automate Tasks in GitHub with GitHub Actions

In this chapter, we will focus on learning how to automate common tasks in GitHub with GitHub Actions using GitHub Issues. This is often called **IssueOps**. In this chapter, we'll create a simple solution that allows you to manage repositories. This does not make sense for a personal account, but the solution should be easily adaptable for an enterprise context in which governance of repositories – such as naming conventions and permissions – is an important topic. The chapter contains the following recipes:

- Creating an issue template
- Using the GitHub CLI and GITHUB_TOKEN to access resources
- Using environments for approvals and checks
- Reusable workflows and composite actions

Technical requirements

If you want to follow along with all the details, you will need a GitHub organization. You can create one in GitHub for free. You can also just use your personal account – it will work the same, but it is less of a real-world scenario. You can author the workflows in Visual Studio Code or in the browser – whatever feels right for you.

Creating an issue template

In this recipe, you will create a simple issue template that you can later extend to gather user input for your IssueOps workflows.

Getting ready...

We'll add the issue template to the repository that you have used in previous chapters. You can clone the repository locally and work in Visual Studio Code or you can do this part in the browser – it doesn't matter. You can follow the examples in my repository (`https://github.com/wulfland/GitHubActionsCookbook`).

How to do it...

1. Create a new file called `.github/ISSUE_TEMPLATE/repo_request.yml` in the repository. GitHub will automatically treat the file in the `.github/ISSUE_TEMPLATE` folder as an issue template as long as it is a YAML or Markdown file.

2. Add a name and description for the template:

    ```
    name: '⬚   Repository Request'
    description: 'Request a new repository.'
    ```

3. Prefill the title of the new issue with a default value:

    ```
    title: '⬚   Repository Request: '
    ```

4. Apply one or multiple labels to the new issue:

    ```
    labels:
      - 'repo-request'
      - 'issue-ops'
    ```

    Note that these labels must exist in the repository. Create them, in case they are not available (use `gh label list` to check):

    ```
    $ gh label create repo-request
    $ gh label create issue-ops
    ```

    You can also provide a description or an explicit color string if you want:

    ```
    $ gh label create repo-request \
    > -c=#D541D0 \
    > -d="Request a new repository"
    ```

5. Assign the issue to one or more users or teams. Just use your GitHub handle in this case:

    ```
    assignees:
      - wulfland
    ```

6. You can automatically assign a new issue to a GitHub project. The syntax is `{owner}/{project id}`:

    ```
    projects: 'wulfland/19'
    ```

Note that the person who creates the issue needs `write` access to that project.

If you don't have a project, then just create a new one. Click the + icon in the top-right corner and select **New project** (see *Figure 5.1*).

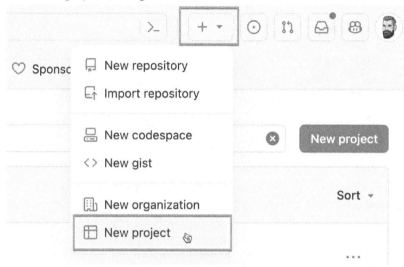

Figure 5.1 – Creating a new project

Pick a template for the project or start from scratch. For a simple project to manage repository requests, you can just start with a simple **Board** (see *Figure 5.2*).

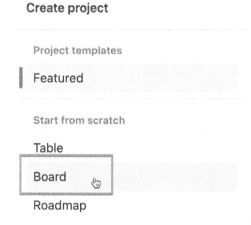

Figure 5.2 – Pick a template or start from scratch

Give the project a name – i.e., `Repository Requests`, and click **Create project**. Get the project ID from the URL of the project: `https://github.com/users/{owner}/projects/{id}`.

7. In the body of the form, you can define different fields. Start with a simple text input for the name of the repository that gets requested. You can make fields required and add additional labels, default values, or placeholders:

```
body:
  - type: input
    id: name
    attributes:
      label: 'Name'
      description: 'Name of the repository in lower-case and
kebab casing.'
      placeholder: 'your-name-kebab'
    validations:
      required: true
```

8. Normally, you'll have a department, region, or team that you would pick for permissions or naming conventions. Add a simple dropdown with two sample departments to pick from:

```
  - type: dropdown
    id: department
    attributes:
      label: 'Department'
      description: 'Pick your department. It will be used as a
prefix for the repository name.'
      multiple: false
      options:
        - dep1
        - dep2
      default: 0
    validations:
      required: true
```

9. Commit the file to your repository.

10. Under **Issues** | **New issue**, you can now pick your template and click **Get started** (see *Figure 5.3*).

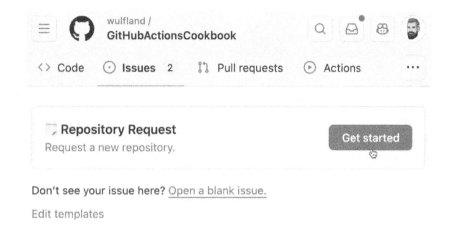

Figure 5.3 – Creating an issue from a template

11. Note that the labels, projects, and assignees are set automatically, and the controls are rendered as required fields and set with the correct defaults (see *Figure 5.4*).

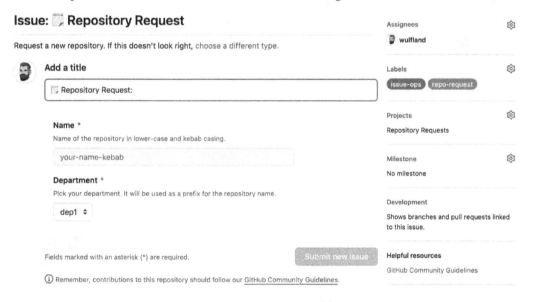

Figure 5.4 – Using the issue template to create a new issue

Fill out the form and save the new issue.

How it works...

Issue and pull request templates are a powerful tool to guide users when creating issues or pull requests. You can generate templates through the UI to make them more discoverable (see `https://docs.github.com/en/communities/using-templates-to-encourage-useful-issues-and-pull-requests/syntax-for-issue-forms`). The templates can be pure Markdown, but with the new custom templates we are using in this recipe, you can create rich forms with multiple form elements such as `markdown`, `textarea`, `input`, `dropdown`, and `checkbox`. You can also add validation and provide default values. For the complete syntax for the GitHub form schema, please refer to `https://docs.github.com/en/communities/using-templates-to-encourage-useful-issues-and-pull-requests/syntax-for-githubs-form-schema`. Note that after filling out the form, the data is added to the body of the issue or pull request as `markdown`. Templates only support the user at the time they create the issue or pull request. After that, it is just `markdown` when editing it.

There is more...

GitHub will display all valid Markdown or YAML form templates from the `.github/ISSUE_TEMPLATE` folder when creating a new issue. But you can configure additional links to external systems, and you can configure whether you want to allow blank issues or force the user to pick a template (see *Figure 5.5*).

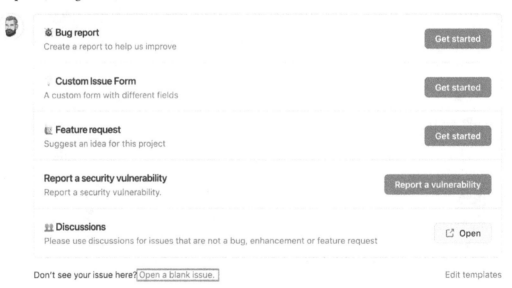

Figure 5.5 – Configuring the template picker form

To configure the template picker, add a `config.yml` file to the `.github/ISSUE_TEMPLATE` folder. Set `blank_issues_enabled` to `true` or `false` and add additional links to the `contact_links` array:

```
blank_issues_enabled: false
contact_links:
  - name: 😺 Discussions
    url:  https://github.com/wulfland/AccelerateDevOps/discussions/new
    about: Please use discussions for issues that are not a bug,
enhancement or feature request.
```

See the documentation (`https://docs.github.com/en/communities/using-templates-to-encourage-useful-issues-and-pull-requests/configuring-issue-templates-for-your-repository`) for more information.

Using the GitHub CLI and GITHUB_TOKEN to access resources

In this recipe, we will parse the issue from the previous chapter and interact with the issue using the GitHub CLI in the workflow.

Getting ready...

You will need the issue template from the previous chapter. You can create the workflow either in Visual Studio Code or directly in GitHub.

How to do it...

1. Create a new workflow `.github/workflows/issue-ops.yml` and name it `issue-ops`:

    ```
    # Issue ops
    name: issue-ops
    ```

2. Use the `issues` trigger for the workflow. Note that we are not using the `created` or `edited` events but rather `labeled`. This allows users to relabel issues when modifying the request:

    ```
    on:
      issues:
        types: [labeled]
    ```

3. Add an `issue-ops` job:

    ```
    jobs:
      issue-ops:
    ```

We only want to run this job for specific labels. Add a condition like the following to the job:

```
if: ${{ github.event.label.name == 'repo-request' }}
```

The job can run on the latest Ubuntu image:

```
runs-on: ubuntu-latest
```

4. To interact with the issues, the workflow needs `write` permissions on `issues`. To use the GitHub CLI, it also needs `read` permissions for `contents`:

```
permissions:
    issues: write
    contents: read
```

5. There is an action in the marketplace that can help with parsing the body of issues created as forms. Add `zentered/issue-forms-body-parser` and give it an `id` property to later access the output:

```
steps:
- name: Issue Forms Body Parser
  id: parse
  uses: zentered/issue-forms-body-parser@v2.0.0
```

6. Next, add our main script. Set `id` to access the output on the job level. Also, set the `GH_TOKEN` environment variable to `GITHUB_TOKEN`. This token will be used by the CLI to interact with the issue:

```
- name: Repository Request Validation
  id: repo-request
  env:
    GH_TOKEN: ${{ github.token }}
  run: |
```

As a first step, read the values from `name` and `department` from the output of the step with the `parse` ID using `jq` and store them in a variable. In JavaScript, we could directly access the object, but in a bash script, we have to use `jq`:

```
repo_name=$(echo '${{ steps.parse.outputs.data }}' | jq -r
'.name.text')
repo_dept=$(echo '${{ steps.parse.outputs.data }}' | jq -r
'.department.text')
```

Combine both variables with your final name (in our case, the department is the prefix):

```
repo_full_name=$repo_dept-$repo_name
```

Set the name as the output to be used in later steps or jobs:

```
echo "REPO_NAME=$repo_full_name" >> "$GITHUB_OUTPUT"
```

7. We are going to add some validation logic. I'll add two examples here. You can look up the rest in the workflow file: `https://github.com/wulfland/GitHubActionsCookbook/blob/main/.github/workflows/issue-ops.yml`.

    First, set the default message and exit code. The default message consists of two parts: first, we want to mention the user who has created the issue, and then we add the message depending on the output of the validation:

    ```
    mention="@${{ github.event.issue.user.login }}: "
    message="Requested repository '$repo_full_name' will be sent for
    approval."
    exitcode=0
    ```

    Next, add a validation rule that the name cannot be empty:

    ```
    # shall not be empty
    if [ -z $repo_full_name ]; then
      message="Repository name is empty.";
      exitcode=1;
    fi;
    ```

    Also, add a validation that it can only use alphanumeric characters and the minus sign (if you want to use kebab casing for your names):

    ```
    # shall be alphanumeric and minus only
    if [[ "$repo_full_name" =~ [^\-a-zA-Z0-9] ]]; then
      message="Repository name shall be alphanumeric and minus
    only.";
      exitcode=1;
    fi;
    ```

8. In case the validation fails, remove the label from the issue and tell the user to fix the issue and reapply the label:

    ```
    if [ $exitcode -ne 0 ]; then
      gh issue edit ${{ github.event.issue.number }} \
        --remove-label repo-request
      message=$message" Please fix the issue and try again by
    applying the label 'repo-request' again to the issue.";
    fi;
    ```

9. Finally, comment the message on the issue and fail the job in case the validation has failed:

    ```
    gh issue comment ${{ github.event.issue.number }} \
      -b „$mention $message"
    exit $exitcode
    ```

10. Set the REPO_NAME output to the output set in the step. We will use this in the next job to create the actual repository:

```
outputs:
  REPO_NAME: ${{ steps.repo-request.outputs.REPO_NAME }}
```

11. Now, create a new issue using the form template. Start with an invalid name (i.e., my_repo) and see how it adds the comment to the issue (see *Figure 5.6*).

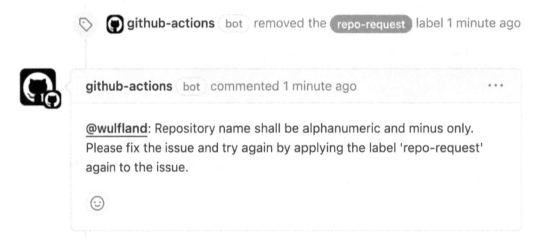

Figure 5.6 – Interacting with the issue from the workflow

Fix the name (to i.e., my-repo) and apply the repo-request label again to the issue.

How it works...

Let's understand the workflow behind it.

Workflow permissions and the GITHUB_TOKEN

At the start of each workflow job, GitHub automatically creates a unique GITHUB_TOKEN that you can use in your workflow to interact with GitHub. You can use this token to authenticate in the workflow job.

You can configure the default permissions for personal accounts and for organizations and repositories – the default is read-only. The best practice is to leave this at read-only and grant explicit permissions in workflows or jobs.

The permissions for GITHUB_TOKEN can be configured either as a top-level key, to apply to all jobs in the workflow, or within specific jobs. When you add the permissions key within a specific job, all actions and run commands within that job that use GITHUB_TOKEN gain the access rights you specify.

For each of the available scopes, you can assign one of the permissions: `read`, `write`, or none.

> **Note**
>
> If you specify the access for any of these scopes, all of those that are not specified are automatically set to none!

You can also set all permissions at once. The following will set all permissions to `read-only`:

```
permissions: read-all
```

The following will grant write access to all scopes:

```
permissions: write-all
```

The last one will set all scopes to none:

```
permissions: {}
```

In our example, we need permission to write to issues and the CLI needs read access to the repository:

```
permissions:
    issues: write
    contents: read
```

All others will be automatically set to none.

See `https://docs.github.com/en/actions/using-jobs/assigning-permissions-to-jobs` for more information on `GITHUB_TOKEN` and workflow permissions.

Step and job outputs

In *Chapter 3*, *Building GitHub Actions*, you learned about **environment files**. We use them in this recipe to set the output that we'll use in subsequent jobs. We also use them to access the data from the form parser.

The form parser is an action from the marketplace that helps us access the body of an issue that was created with an issue template. We use `id` to access the output:

```
steps:
- name: Issue Forms Body Parser
  id: parse
  uses: zentered/issue-forms-body-parser@v2.0.0
```

In JavaScript (i.e., the GitHub Script action), you could directly access the JSON objects:

```
console.log(data.name.text);
console.log(data.department.text);
```

In bash, you can't directly access JSON properties like you can in JavaScript. You need to use a command-line JSON processor such as jq to parse the JSON string and access its properties:

```
repo_name=$(echo '${{ steps.parse.outputs.data }}' | jq -r '.name.
text')
```

Commenting on issues using the GitHub CLI

In *Chapter 3*, you commented on an issue using octokit and the REST API. In this recipe, we use the GitHub CLI to do this. For the CLI to work, you have to first check out the repository. You also have to set the GH_TOKEN environment variable to the workflow token on the workflow step:

```
env:
  GH_TOKEN: ${{ github.token }}
```

Using the CLI is easy – and we leverage the fact that we can use @ plus the username to mention the user in the comment:

```
mention="@${{ github.event.issue.user.login }}: "
message="Requested repository '$repo_full_name' will be sent for
approval."

gh issue comment ${{ github.event.issue.number }} \
  -b "$mention $message"
```

Using environments for approvals and checks

In this recipe, we are going to use environment approvals to acquire approval before creating the repository. We will also use a GitHub App to authenticate as the repository creation normally happens in the organization scope and cannot be done with GITHUB_TOKEN. You must either use a **GitHub App** or a **personal access token** (**PAT**) with the right scopes.

Getting ready...

Make sure you have completed the previous recipe and continue in the same repository.

How to do it...

1. In your repository, go to **Setting | Environments** and click **New environment**.

2. Name the environment repo-creation and click **Configure environment**.

3. Add yourself as **Required reviewer** and don't allow administrators to bypass the rule (see *Figure 5.7*).

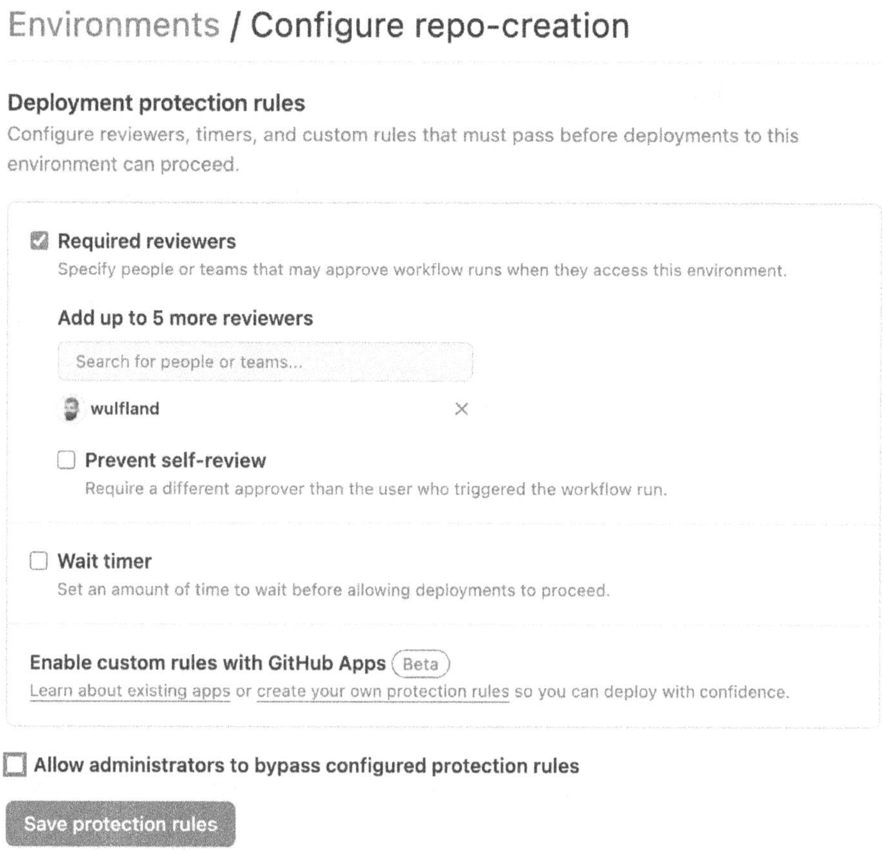

Environments / Configure repo-creation

Deployment protection rules

Configure reviewers, timers, and custom rules that must pass before deployments to this environment can proceed.

> ☑ **Required reviewers**
> Specify people or teams that may approve workflow runs when they access this environment.
>
> **Add up to 5 more reviewers**
>
> | Search for people or teams... |
>
> 🧔 **wulfland**   ✕
>
> ☐ **Prevent self-review**
> Require a different approver than the user who triggered the workflow run.
>
> ---
>
> ☐ **Wait timer**
> Set an amount of time to wait before allowing deployments to proceed.
>
> ---
>
> **Enable custom rules with GitHub Apps** (Beta)
> Learn about existing apps or create your own protection rules so you can deploy with confidence.

☐ **Allow administrators to bypass configured protection rules**

[Save protection rules]

Figure 5.7 – Configuring the environment protection rules

4. Click **Save protection rules**.

5. The next step is to create an app to authenticate. I recommend using an organization to try this recipe – but you could also use your personal account. Go to `https://github.com/settings/apps` and click **New GitHub App**.

6. Give it a unique name (i.e., `{your username}-repo-creation`) and set **Homepage URL** to the URL of your repository.

7. Under **Repository permissions**, select **Administration: Read and write**, **Contents: Read and write**, **Issues: Read and write**, and under **Organization permissions**, select **Administration: read and write**.

8. Save the app. In your newly created app, click on **Generate a private key**. The private key will be automatically downloaded.

9.   Copy **App ID** from the **General** tab of the app.

10. In the app in GitHub, select **Install App** and click **Install**. Pick your organization or account, click **install**, leave **All repositories** selected, and click **Install**.

11. Go back to the environment in your repository. Add a new secret PRIVATE_KEY and add the content of the key file you downloaded earlier. Also, add an APP_ID variable with the ID of your app (see *Figure 5.8*).

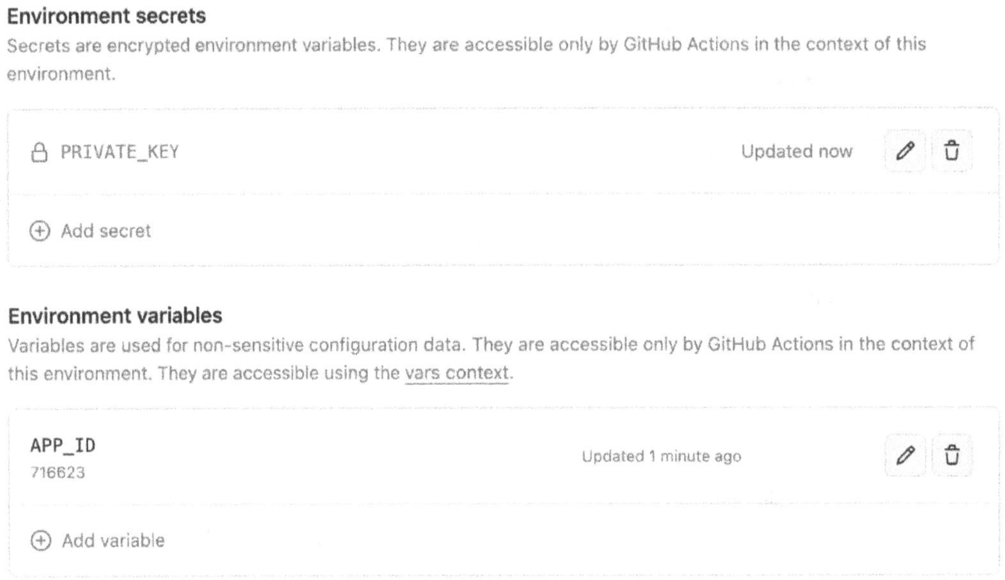

Environment secrets

Secrets are encrypted environment variables. They are accessible only by GitHub Actions in the context of this environment.

🔒 PRIVATE_KEY	Updated now	✏️ 🗑️
⊕ Add secret		

Environment variables

Variables are used for non-sensitive configuration data. They are accessible only by GitHub Actions in the context of this environment. They are accessible using the vars context.

APP_ID 716623	Updated 1 minute ago	✏️ 🗑️
⊕ Add variable		

Figure 5.8 – Adding variables and secrets to an environment

12. Edit your workflow file and add an additional create-repo job. Make the job depend on the previous job (needs: issue-ops) and assign it the environment that we have created earlier (environment: repo-creation):

```
create-repo:
  needs: issue-ops
  runs-on: ubuntu-latest
  environment: repo-creation
```

13. To interact with the issue using the GitHub token, set the permission for the workflow:

```
permissions:
  issues: write
  contents: read
```

14. We can use global environment variables that we can use in all steps. Set `REPO_OWNER` to the organization or account in which you have installed the app. You could also save the value as a variable in the environment. Set `REPO_NAME` to the output of the previous job. `USER` and `ISSUE_NUMBER` are set to the values of the context for easy access:

```
env:
    REPO_OWNER: ${{ vars.ORGANIZATION }}
    REPO_NAME: ${{ needs.issue-ops.outputs.REPO_NAME }}
    USER: ${{ github.event.issue.user.login }}
    ISSUE_NUMBER: ${{ github.event.issue.number }}
```

15. To authenticate using the app, we can use the `actions/create-github-app-token` action. Give it an ID to reference it later:

```
steps:
    - name: Create app token
      uses: actions/create-github-app-token@v1.6.2
      id: get-workflow-token
      with:
        app-id: ${{ vars.APP_ID }}
        private-key: ${{ secrets.PRIVATE_KEY }}
        owner: ${{ vars.ORGANIZATION }}
```

16. Create the repository and set the URL as an output parameter:

```
    - name: Create repository
      id: create-repo
      env:
        GH_TOKEN: ${{ steps.get-workflow-token.outputs.token }}
      run: |
        REPO_URL=$(gh repo create $REPO_OWNER/$REPO_NAME --private
--clone)
        echo "repo_url=$REPO_URL" >> "$GITHUB_OUTPUT"
        echo "Repositeory '$REPO_NAME' has been successully created:
$REPO_URL"
```

17. If the operation is successful, comment on the issue to inform the user that the repository has been created:

```
    - name: Notify User
      if: ${{ success() }}
      env:
        GH_TOKEN: ${{ github.token }}
        REPO_URL: ${{ steps.create-repo.outputs.repo_url }}
      run: |
        gh issue comment $ISSUE_NUMBER \
```

```
        -b "@$USER: Repository '$REPO_OWNER/$REPO_NAME' has been
created successfully: $REPO_URL" \
        --repo ${{ github.event.repository.full_name }}
```

18. In case of an error, also inform the user by commenting on the issue:

```
- name: Handle Exception
  if: ${{ failure() }}
  env:
    GH_TOKEN: ${{ github.token }}
  run: |
    gh issue comment $ISSUE_NUMBER \
      -b "@$USER: Repository '$REPO_OWNER/$REPO_NAME' creation
failed. Please contact the administrator."\
        --repo ${{ github.event.repository.full_name }}
```

19. Commit and push the workflow and create a new **Repository Request** issue. You will receive a notification according to your notification settings to review the workflow as you are configured as the required reviewer for the environment. The workflow looks like *Figure 5.9*.

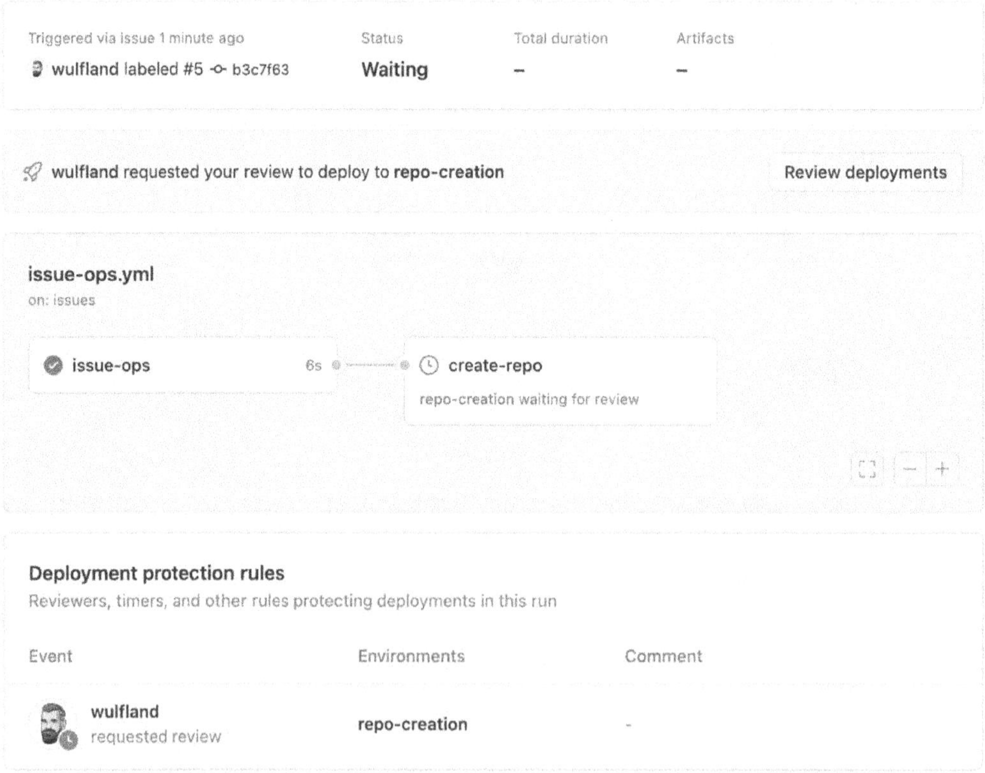

Figure 5.9 – Using environments for manual workflow approvals

20. Click **Review deployments**, select the environment, and click **Approve and deploy** (see *Figure 5.10*).

Figure 5.10 – Approving deployment to an environment

The workflow will create the repository and inform the user who initiated the request using comments in the issue (see *Figure 5.11*).

Figure 5.11 – The workflow notifying the user about successful
repository creation using the issue comments

Check that the link works and the repository is created correctly.

How it works...

You can assign jobs in a workflow to environments, allowing you to add protection rules and specific variables and secrets for that environment.

Environments

Environments are created in a repository using the web UI or the API:

```
curl -L \
  -X PUT \
  -H "Accept: application/vnd.github+json" \
  -H "Authorization: Bearer <YOUR-TOKEN>" \
  -H "X-GitHub-Api-Version: 2022-11-28" \
  https://api.github.com/repos/OWNER/REPO/environments/<NAME> \
  -d '{"wait_timer":30,"prevent_self_review":false,"reviewers":[{"-
type":"User","id":1},{"type":"Team","id":1}],"deployment_branch_poli-
cy":{"protected_branches":false,"custom_branch_policies":true}}'
```

In most cases, you will create it using the web UI like we did in this recipe. In the workflow, the environment is referenced using its name:

```
jobs:
  deployment:
    runs-on: ubuntu-latest
    environment: production
```

You can add an additional URL that will be displayed in the workflow:

```
environment:
  name: production
  url: https://writeabout.net
```

This allows you to create dynamic environments in the workflow and deploy to them. This is useful if you want to deploy every pull request to an isolated environment to test it.

You can protect environments with different protection rules:

- **Required reviewers**: You can list up to six users or teams as required reviewers to approve workflow jobs that reference the environment. The reviewers must have at least read access to the repository. Only one of the required reviewers needs to approve the job for it to proceed.

- **Wait timer**: You can pause a workflow for a specific amount of time after the job is initially triggered. The time (in minutes) must be an integer between 0 and 43,200 (30 days). You could use the API to cancel the workflow in that time.

- **Deployment branches and tags**: Use deployment branches and tags to restrict which branches and tags can deploy to the environment. The options for deployment branches and tags for an environment are **No restricted**, **Protected branches only**, or **Selected branches and tags**. In the letter settings, you can add name patterns to target individual or groups of branches or tags – such as `main` or `release/*`. Connecting environments to branch protection rules is very powerful, as you have a lot more protection rules for that – such as enforcing code owners or deployment to specific environments.

Environments also have specific secrets and variables that allow you to use different configurations in the same workflow.

There is also **Custom deployment rules** to protect your environments. This feature is still in public beta at the time of writing this book. Custom deployment rules are basically GitHub Apps that allow you to write your own integration. This allows services such as Datadog, Honeycomb, and ServiceNow to provide automated approvals for deployments.

To learn more about environments, see `https://docs.github.com/en/actions/deployment/targeting-different-environments/using-environments-for-deployment`.

Authentication

You can do a lot with the GitHub token and workflow permissions. But especially when automating things on the organization level, you probably need to use either a PAT or a GitHub App. GitHub Apps are the recommended way as they are not tied to a user. You already learned about GitHub Apps in the previous chapter. To use GitHub Apps to authenticate in a workflow, you can use the `actions/create-github-app-token` action:

```
steps:
  - name: Create app token
    uses: actions/create-github-app-token@v1.6.2
    id: get-workflow-token
    with:
      app-id: ${{ vars.APP_ID }}
      private-key: ${{ secrets.PRIVATE_KEY }}
```

It needs the app ID and private key that we store as environment variables and secrets. The token can then be accessed using the output of the workflow step:

```
  - name: Create repository
    env:
      GH_TOKEN: ${{ steps.get-workflow-token.outputs.token }}
```

With this action, it is really easy to use GitHub Apps in your workflow.

There is more...

Issues are a great way to interact with your users – but you probably also want to store the state of your automation in another place. You could update a YAML or markdown file or call an external system. However, you can also use GitHub Projects to visualize the issues. This way, the issue represents the state over the life cycle of an automated object.

GitHub Projects is very flexible, and you can have issues and pull requests from different repositories in it. This also means it is quite complex and you will need internal IDs to reference the fields. Before adjusting your workflow, run the following command for the project you are tracking your issues at (in my case, the ID is 19 and the owner is wulfland):

```
$ gh project field-list <ID> --owner <OWNER> --format json | jq
```

This will give you a JSON object with all the fields in your project. Look up the IDs. For the Status field, you will also need the options ID:

```
{
        "id": "PVTSSF_1AHOAFCCsc4AZoDtzgQZkEg",
        "name": "Status",
        "type": "ProjectV2SingleSelectField",
        "options": [
          {
            "id": "f75ad846",
            "name": "Request"
          },
          {
            "id": "e05aa0a3",
            "name": "Repository Created"
          },
          {
            "id": "98236657",
            "name": "Deleted"
          }
        ]
      },
```

Store the internal IDs as variables in the environment as displayed in *Figure 5.12*.

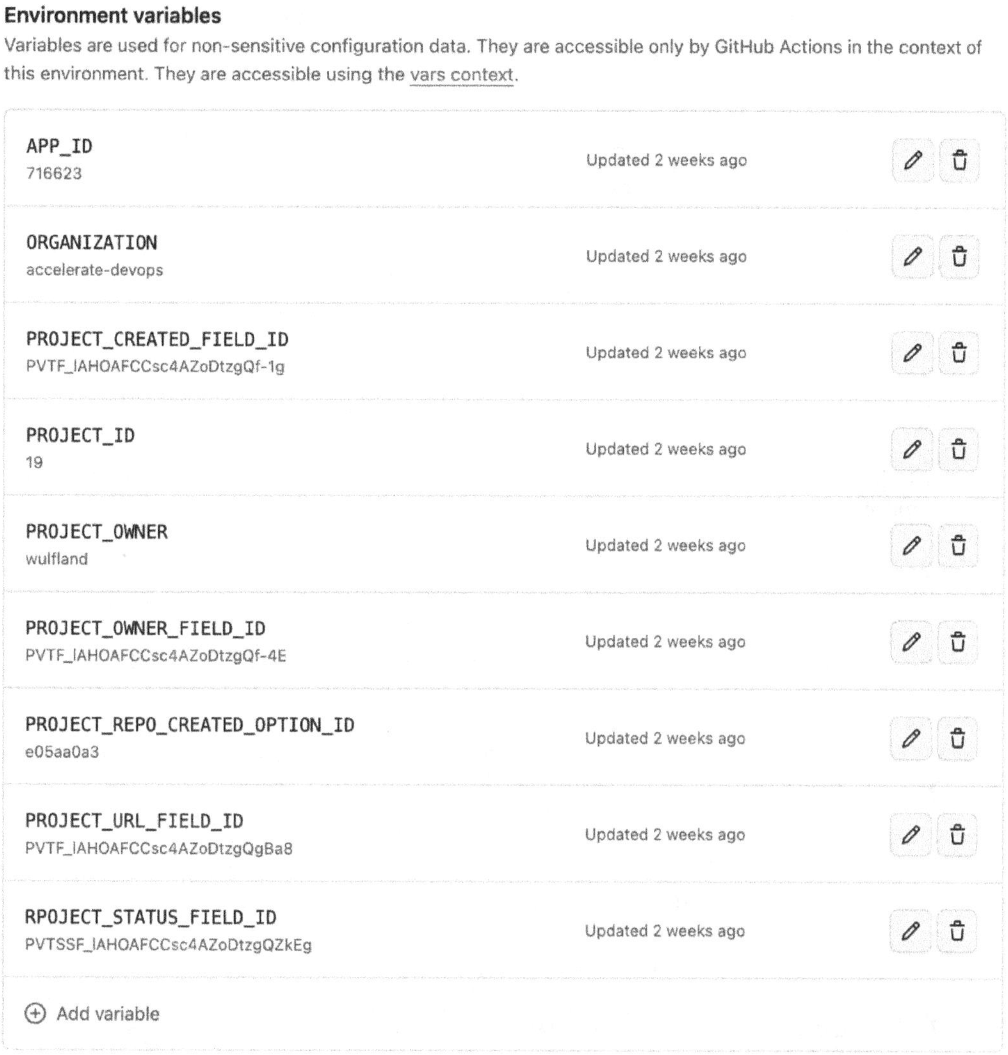

Environment variables

Variables are used for non-sensitive configuration data. They are accessible only by GitHub Actions in the context of this environment. They are accessible using the vars context.

APP_ID 716623	Updated 2 weeks ago	✏️ 🗑️
ORGANIZATION accelerate-devops	Updated 2 weeks ago	✏️ 🗑️
PROJECT_CREATED_FIELD_ID PVTF_IAHOAFCCsc4AZoDtzgQf-1g	Updated 2 weeks ago	✏️ 🗑️
PROJECT_ID 19	Updated 2 weeks ago	✏️ 🗑️
PROJECT_OWNER wulfland	Updated 2 weeks ago	✏️ 🗑️
PROJECT_OWNER_FIELD_ID PVTF_IAHOAFCCsc4AZoDtzgQf-4E	Updated 2 weeks ago	✏️ 🗑️
PROJECT_REPO_CREATED_OPTION_ID e05aa0a3	Updated 2 weeks ago	✏️ 🗑️
PROJECT_URL_FIELD_ID PVTF_IAHOAFCCsc4AZoDtzgQgBa8	Updated 2 weeks ago	✏️ 🗑️
RPOJECT_STATUS_FIELD_ID PVTSSF_IAHOAFCCsc4AZoDtzgQZkEg	Updated 2 weeks ago	✏️ 🗑️

⊕ Add variable

Figure 5.12 – Using the internal project IDs to reference fields and options

In the workflow, add an additional step and set the environment variables:

```
- name: Update Project
  if: ${{ success() }}
  env:
    GH_TOKEN: ${{ secrets.PROJECT_TOKEN }}
    REPO_URL: ${{ steps.create-repo.outputs.repo_url }}
    PROJECTNUMBER: ${{ vars.PROJECT_ID }}
    PROJECTOWNER: ${{ vars.PROJECT_OWNER}}
```

First, you have to receive the internal project ID as this is required in subsequent commands. The normal project ID – the number – cannot be used in all commands:

```
run: |
    project_id=$(gh project list --owner "$PROJECTOWNER" --format json
| jq -r '.projects[] | select(.number=='$PROJECTNUMBER') | .id')
```

Next, get the internal issue ID:

```
issue_id=$(gh project item-list $PROJECTNUMBER \
  --owner "$PROJECTOWNER" \
  --format json \
  | jq -r '.items[] \
  | select(.content.number=='$ISSUE_NUMBER') | .id')
```

Now, you can update the fields to the values from the workflow. Set the status of the item to the `created` option:

```
gh project item-edit \
  --id $issue_id \
  --field-id ${{ vars.RPOJECT_STATUS_FIELD_ID }} \
  --single-select-option-id ${{ vars.PROJECT_REPO_CREATED_OPTION_ID }}
\
  --project-id $project_id
```

Set the URL of the repository created to the URL field:

```
gh project item-edit \
  --id $issue_id \
  --field-id ${{ vars.PROJECT_URL_FIELD_ID }} \
  --text $REPO_URL \
  --project-id $project_id
```

And, set the created date field to the current date:

```
gh project item-edit --id $issue_id \
  --field-id ${{ vars.PROJECT_CREATED_FIELD_ID }} \
  --date $(date +%Y-%m-%d) \
  --project-id $project_id
```

You can now track the status of your repository requests in Projects (*Figure 5.13*).

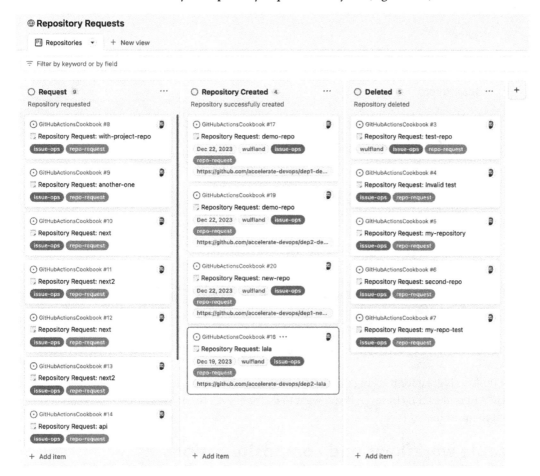

Figure 5.13 – Tracking the state of IssueOps in GitHub Projects

The metadata is also visible on a card in each individual issue (see *Figure 5.14*).

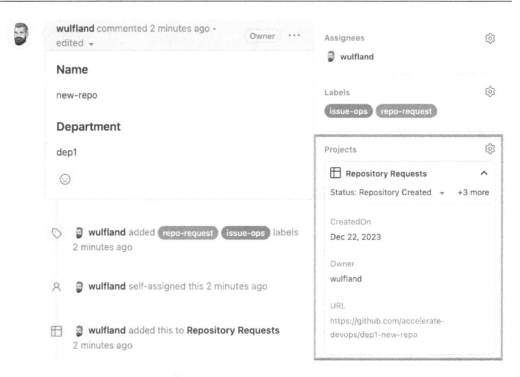

Figure 5.14 – GitHub Projects cards in GitHub Issues

GitHub Projects is great for tracking the status of issues, but it is not so easy to automate it using the GraphQL API or the CLI. However, if you want to do more with IssueOps, it is definitely something you want to invest in.

Reusable workflows and composite actions

If you start automating with IssueOps, workflows get very complex very fast. You will end up with a lot of `if` clauses on jobs and steps. To keep these solutions maintainable, you can use composite actions or reusable workflows to not only reuse functionality but also break down complex workflows into smaller parts.

We have already covered composite actions in previous chapters. In this recipe, we will use a reusable workflow to add the delete functionality to our IssueOps solution.

Getting ready...

Make sure you have completed the previous recipes in this chapter.

How to do it...

1. Create a new `delete-repo.yml` workflow file.

2. As the trigger, we use the `workflow_call` trigger. This indicates that the workflow is a reusable workflow. Define inputs needed by our workflow:

```
on:
  workflow_call:
    inputs:
      REPO_NAME:
        description: 'Repository name'
        required: true
        type: string
      ISSUE_USER:
        description: 'User who created the issue'
        required: true
        type: string
      ISSUE_NUMBER:
        description: 'Issue number'
        required: true
        type: number
```

3. A reusable workflow is just a normal workflow that can have one or multiple jobs that are also associated with an environment:

```
jobs:
  delete:
    runs-on: ubuntu-latest
    environment: repo-cleanup
    steps:
```

You can create a new environment for the deletion of the repository and add `PRIVATE_KEY`, `APP_ID`, `ORGANIZATION`, `PROJECT_OWNER`, and `REPO_OWNER`. Or, you can just reuse the `repo-creation` environment for simplicity.

4. Get the token from the app to authenticate, as we did in the `issue-ops` workflow:

```
- name: Create app token
  uses: actions/create-github-app-token@v1.6.2
  id: get-workflow-token
  with:
    app-id: ${{ vars.APP_ID }}
    private-key: ${{ secrets.PRIVATE_KEY }}
    owner: ${{ vars.ORGANIZATION }}
```

5. Next, delete the repository using the token provided:

```
- name: Delete repository
  id: delete-repo
  env:
    GH_TOKEN: ${{ steps.get-workflow-token.outputs.token }}
    REPO_NAME: ${{ inputs.REPO_NAME }}
    REPO_OWNER: ${{ vars.REPO_OWNER }}
  run: |
      gh repo delete $REPO_OWNER/$REPO_NAME --yes
      echo "Repositeory '$REPO_NAME' has been successully
deleted."
```

6. Notify the user and close the issue:

```
- name: Notify User
  if: ${{ success() }}
  env:
    GH_TOKEN: ${{ github.token }}
    ISSUE_NUMBER: ${{ inputs.ISSUE_NUMBER }}
    ISSUE_USER: ${{ inputs.ISSUE_USER }}
    REPO_NAME: ${{ inputs.REPO_NAME }}
    REPO_OWNER: ${{ vars.REPO_OWNER }}
  run: |
    gh issue comment $ISSUE_NUMBER \
      -b "@$ISSUE_USER: Repository '$REPO_OWNER/$REPO_NAME' has
been deleted successfully." \
      --repo ${{ github.event.repository.full_name }}
    gh issue close $ISSUE_NUMBER \
      --repo ${{ github.event.repository.full_name }}
```

7. In the case of a failure, also notify the user:

```
- name: Handle Exception
  if: ${{ failure() }}
  env:
    GH_TOKEN: ${{ github.token }}
    ISSUE_NUMBER: ${{ inputs.ISSUE_NUMBER }}
    ISSUE_USER: ${{ inputs.ISSUE_USER }}
  run: |
    gh issue comment $ISSUE_NUMBER \
      -b "@$ISSUE_USER: Repository '$REPO_OWNER/$REPO_NAME'
deletion failed. Please contact the administrator." \
      --repo ${{ github.event.repository.full_name }}
```

8. To use this workflow, we create a new workflow file called `handle-issue.yml`. We have it run on labeled issues and grant it write access to issues:

```
name: Handle Issue
on:
  issues:
    types: [labeled]
permissions:
  contents: read
  issues: write
```

9. To parse the issue, we add a common job that uses the same logic we used in the `issue-ops` workflow (just copy it over until we set the output variable):

```
jobs:
  parse-issue:
      runs-on: ubuntu-latest
      outputs:
          REPO_NAME: ${{ steps.repo-request.outputs.REPO_NAME }}
      steps:
      - name: Issue Forms Body Parser
        id: parse
        uses: zentered/issue-forms-body-parser@v2.0.0

      - name: Repository Request Validation
        id: repo-request
        env:
          GH_TOKEN: ${{ github.token }}
        run: |
            repo_name=$(echo '${{ steps.parse.outputs.data }}' |
jq -r '.name.text')
            repo_dept=$(echo '${{ steps.parse.outputs.data }}' |
jq -r '.department.text')
            repo_full_name=$repo_dept-$repo_name

            echo "REPO_NAME=$repo_full_name" >> "$GITHUB_OUTPUT"
```

10. Then, we add the job that will call the other workflow file. Conditionally execute the job when the applied label is `delete-repo`. Pass in the repository name together with the other parameters using the `with` section:

```
repo-deletion:
  name: "Delete a repository"
  if: github.event.label.name == 'delete-repo'
  uses: ./.github/workflows/delete-repo.yml
```

```
needs: parse-issue
with:
  REPO_NAME: ${{ needs.parse-issue.outputs.REPO_NAME }}
  ISSUE_USER: ${{ github.event.issue.user.login }}
  ISSUE_NUMBER: ${{ github.event.issue.number }}
secrets: inherit
```

With `secrets: inherit`, you can allow access to all secrets of the parent workflow without having to specify them all as `secret` parameters.

11. Commit and push your files and apply the `delete-issue` label to the issue that you had used to test the creation of the repositories. Approve the deployment and the repository will be deleted.

Note how jobs from reusable workflows are nested in the workflow jobs section and how they are displayed in the designer (see *Figure 5.15*).

Figure 5.15 – Nested jobs from reusable workflows

Also, note that the issue was closed after the comment that the deletion was successful.

How it works...

Reusable workflows are a great way to structure complex workflows and reuse a more complex functionality that relies on multiple jobs or environments. However, there are some limitations. You can connect up to 4 levels of workflows and you can call a maximum of 20 reusable workflows from a single workflow file. This limit includes any trees of nested reusable workflows that may be called starting from your top-level caller workflow file. Any environment variables set in an `env` context defined at the workflow level in the caller workflow are not propagated to the called workflow. Similarly, environment variables set in the `env` context, defined in the called workflow, are not accessible in the `env` context of the caller workflow. Instead, you must use outputs of the reusable workflow. To reuse variables in multiple workflows, set them at the organization, repository, or environment levels and reference them using the `vars` context.

To learn more about reusable workflows, see `https://docs.github.com/en/actions/using-workflows/reusing-workflows`.

Note that I did not update the project fields after the deletion of the repository. GitHub Projects also supports workflows that can be used to update a field if the issue is closed.

Just navigate in your project to **Workflows**, enable **Item closed**, and set the value of the status field to the desired value (see *Figure 5.16*).

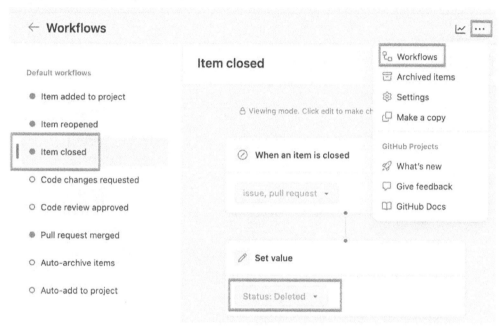

Figure 5.16 – Configuring workflows in GitHub Projects

This will automatically set the status to `Deleted` in our case when the issue is closed.

There is more...

Of course, this is only the starting point. To adopt this in a real-world scenario, you will have to extend the solution to the following:

- The life cycle of teams
- Granting permissions to teams
- Setting a base configuration for the repositories
- Providing different templates for different solution types

But the building blocks are the same.

You also have to refactor the current solution to include the creation of the repository as a reusable workflow and include it in the handle-issue workflow. I left it like this in my repository to keep the complexity of the individual steps as low as possible while providing a real-world solution.

Build and Validate Your Code

In this chapter, you will learn how to use GitHub Actions to build and validate code, validate changes in pull requests, and keep your dependencies up to date. You will learn where to store build output and how to optimize your workflow runs with caching.

We will cover the following recipes:

- Building and testing your code
- Building different versions using a matrix
- Informing the user on details of your build and test results
- Finding security vulnerabilities with CodeQL
- Creating a release and publishing the package
- Versioning your packages
- Generating and using **software bills of materials** (**SBOMs**)
- Using caching in workflows

Technical requirements

For this chapter, you will need an up-to-date version of NodeJS and Visual Studio Code. You can also use GitHub Codespaces.

Building and testing your code

In this recipe, we will create a simple **Continuous Integration** (**CI**) pipeline that builds and validates code and gets integrated into pull request validation. We'll use this code in subsequent recipes – that's why we're using a very simple JavaScript package.

Getting ready

I think it is best to build a new repository and npm package from scratch. Even if you're not familiar with JavaScript, this should not be a challenge. You can compare your code or copy it from the following repository: `https://github.com/wulfland/package-recipe`. If you struggle with this, just clone this repository and work with it.

1. Create a new public repository called `package-recipe`. Initialize it with a `.gitignore` file and a README file and pick `Node` as the template for the `.gitignore` file.

2. Clone your repository locally or open it in Codespaces.

3. Run the following command:

   ```
   $ npm init
   ```

   Follow the wizard. The name of the package is `@<github-user-name>/package-recipe`. As the test command, add the following code:

   ```
   jest && make-coverage-badge
   ```

   Jest is the test framework we will use to test the application, and with `make-coverage-badge`, we will later create a badge that we will add to the README file.

   You can leave the rest of the properties at their default.

4. After the wizard has completed, run the following command:

   ```
   $ npm install
   ```

5. Install the dependencies:

   ```
   $ npm install --save-dev jest
   $ npm install --save-dev make-coverage-badge
   ```

6. Next, let's add the code. Create a new `src/index.js` file and add the following lines:

   ```
   module.exports = function greet () {
      return 'Hello world!'
   }
   ```

   We're creating a simple package that will only return `Hello world!`.

7. Add a new `__tests__/index.test.js` file (with double underscores). Add the following test:

   ```
   describe('index.js', () => {
      it('greet function returns Hello world!', () => {
         const greet = require('../src/index')
   ```

```
        expect(greet()).toBe('Hello world!')
    })
  })
```

This is just a simple test that checks that `index.js` writes the text as expected to the console.

8. Open `package.json` and add the following configuration for Jest:

```
"jest": {
    "verbose": true,
    "coverageReporters": [
      "json-summary",
      [
        "text",
        {
          "file": "coverage.txt",
          "path": "./coverage"
        }
      ],
      "lcov"
    ],
    "collectCoverage": true,
    "collectCoverageFrom": [
      "./src/**"
    ]
  },
```

This will add multiple report outputs for the code coverage.

9. Execute the tests:

```
$ npm run test
```

If everything is set up correctly, this will execute the one test and write reports in a `coverage` folder. It will also generate a badge at `coverage/badge.svg`.

10. Commit your files and push them to GitHub.

11. In the repository, go to **Settings** and scroll down to **Pull Requests**. Check **Allow auto-merge** and **Automatically delete head branches**.

How to do it...

Now that we have our package repository ready that we are going to use for the next recipes, let's start with adding a CI workflow:

1. Create a new `_github/workflows/ci.yml` file.

Name the workflow CI and trigger it on every pull request to validate the changes. Also, trigger it on every push to the main branch to build a new version that can be released:

```
name: CI

on:
    pull_request:
    push:
      branches:
        - main
```

2. Add a build job that runs on ubuntu-latest and checks out of the repository:

```
jobs:
  build:
    runs-on: ubuntu-latest

    steps:
      - uses: actions/checkout@v4
```

3. Configure the workflow runner to target a specific NodeJS version. We can use a wildcard for minor versions and have the action use the latest version available:

```
- uses: actions/setup-node@v4
  with:
    node-version: "21.x"
    check-latest: true
```

4. Build and test the code:

```
- name: Install dependencies
  run: npm install

- name: Run tests
  run: npm test
```

5. Commit and push the workflow to the main branch. This should already trigger the workflow because of the push trigger, and the build and test should succeed.

6. To test the validation, create a new branch:

```
$ git switch -c fail-pr
```

7. Modify index.js to something that will fail the test (that is, console.log('Hello Mars!');). Add and commit the changes and create a pull request:

```
$ git add index.js
$ git commit
```

```
$ git push -u origin fail-pr
$ gh pr create --fill
```

The pull request will trigger the workflow, and it will fail because the npm run test command will return a nonzero value.

How it works...

Let's understand how the validation works.

Checkout

The first thing necessary for CI is checking out of your repository with the checkout action (https://github.com/actions/checkout). The action has many parameters. For example, you can set the depth of the git history you pull down to the runner. The default is 1, and that will only download the HEAD branch. In the recipe to generate version numbers from git, we'll set that to 0 to download all branches and tags:

```
steps:
  - uses: actions/checkout@v4
    with:
      fetch-depth:0 # Default: 1
```

Other options are lfs to determine if Git **Large File Storage** (LFS) files should be downloaded or not (default is false) or submodules:

```
steps:
  - uses: actions/checkout@v4
    with:
      lfs: true # Default: false
      submodules: true # Default: false
```

For large monorepos, you can also perform a sparse checkout and only get data for a specific area of the repo. This example only checks out the .github, src, and __tests__ folders:

```
steps:
  - uses: actions/checkout@v4
    with:
      sparse-checkout: |
        .github
        src
        __tests__
```

This can reduce the time and storage needed by the workflow tremendously.

Setup environment

There are different setup actions for the following programming languages:

- Node
- Python
- Java
- Go
- .NET
- Ruby
- Elixir
- Haskell

In our case, we use `setup-node` (`https://github.com/actions/setup-node`). Setup actions ensure the correct binaries and environment variables are set in the build process. All the actions take some form of version parameter:

```
- uses: actions/setup-node@v4
  with:
    node-version: 21
```

Wildcards are supported as well as aliases. Examples are `21.x`, `21.5.0`, `>=21.5.0`, `lts/Hydrogen`, `21-nightly`, `latest`. If you use wildcards, then you can set `check-latest: true` to check for the latest version available.

Most checkout actions also support different registries to download dependencies. For node, the parameter is `registry-url`, which will use the URL provided and a token from `env.NODE_AUTH_TOKEN` to connect to the specific registry.

Checkout actions normally also cache dependencies. In many cases, it is more efficient to use the corresponding setup action than implementing caching yourself (we have a recipe for that later in this chapter).

There's more...

A validation workflow is best combined with branch protection and rulesets – and you can add a linter to the mix to further increase the quality.

super-linter

In *Chapter 2, Authoring and Debugging Workflows*, we used a linter to validate and annotate our workflows in pull requests. The same can be done for code. There is a GitHub action called `super-linter` (see `https://github.com/super-linter/super-linter`) that basically combines all available linters in one action. You can use it to lint your code like this:

```
permissions:
  contents: read
  packages: read
  # To report GitHub Actions status checks
  statuses: write

steps:
  - name: Checkout code
    uses: actions/checkout@v4
    with:
      # super-linter needs the full git history to get the
      # list of files that changed across commits
      fetch-depth: 0

  - name: Super-linter
    uses: super-linter/super-linter@v5.7.2
    env:
      DEFAULT_BRANCH: main
      # To report GitHub Actions status checks
      GITHUB_TOKEN: ${{ secrets.GITHUB_TOKEN }}
```

The image of `super-linter` is quite huge and can impact the performance of your workflows. If you don't need all languages, then you can also use the slim version:

```
super-linter/super-linter/slim@[VERSION]
```

It excludes linters for `rust`, `dotenv`, `armttk`, `pwsh`, and `c#`, and the image size is much smaller. In `v5`, this will reduce the size from 7.35 GB to 4.88 GB in `slim-v5`, and this will reduce the time to just pull down the image from about 2 minutes to about 1 minute.

Branch protection

In *Chapter 2, Authoring and Debugging Workflows*, I also introduced you to branch protection. You can protect one or multiple branches in a repository with rules and enforce that commits are merged with a pull request and certain checks are successful. In addition to manual reviews, you can request status checks to be successful (see *Figure 6.1*):

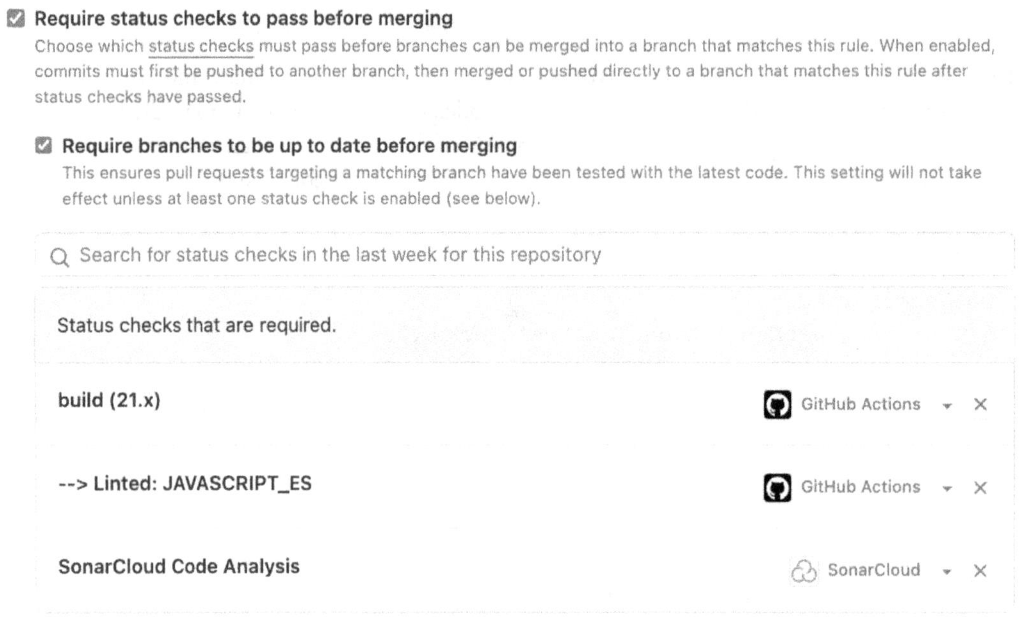

☑ **Require status checks to pass before merging**

Choose which status checks must pass before branches can be merged into a branch that matches this rule. When enabled, commits must first be pushed to another branch, then merged or pushed directly to a branch that matches this rule after status checks have passed.

☑ **Require branches to be up to date before merging**

This ensures pull requests targeting a matching branch have been tested with the latest code. This setting will not take effect unless at least one status check is enabled (see below).

Q Search for status checks in the last week for this repository

Status checks that are required.

build (21.x)   GitHub Actions ▾ ✕

--> Linted: JAVASCRIPT_ES   GitHub Actions ▾ ✕

SonarCloud Code Analysis   SonarCloud ▾ ✕

Figure 6.1 – Adding GitHub Action workflow jobs as status checks for a protected branch

Status checks can be a single job of a workflow – or any integration that uses the `status` API to report status. You can enable a policy for `main` in your repository for the next recipes and set the status check policy to the jobs of the workflow. See `https://docs.github.com/en/ repositories/configuring-branches-and-merges-in-your-repository/ managing-protected-branches/about-protected-branches` for more information on protected branches.

Rulesets

Rulesets are the more powerful successor of branch protection. They have the same power as protected branches. You can also define status checks the same way you do in protected branches (see *Figure 6.2*):

☑ **Require status checks to pass**
Choose which status checks must pass before the ref is updated. When enabled, commits must first be pushed to another ref where the checks pass.

Additional settings ∧

☐ **Require branches to be up to date before merging**
Whether pull requests targeting a matching branch must be tested with the latest code. This setting will not take effect unless at least one status check is enabled.

Status checks that are required		+ Add checks ▾
SonarCloud Code Analysis	☁ SonarCloud	🗑
--> Linted: JAVASCRIPT_ES	⊙ GitHub Actions	🗑
build (21.x)	⊙ GitHub Actions	🗑

Figure 6.2 – Adding GitHub Action workflow jobs as status checks in a ruleset

Most of the rules that can be used in rulesets are similar to branch protection rules and can be used in combination without changing existing ones.

Rulesets are more advanced and have the following advantages over protection rules:

- Rulesets can be defined at the organization level and can target multiple repositories.
- Rules can be layered by applying multiple rulesets at the same time.
- Rulesets have a status that allows you to activate and deactivate rulesets in a repository without the need to delete them.
- You only need read access to a repository to view all active rulesets. This makes it much more transparent to contributors which rules apply.
- There are some additional rules that are not available in protection rules, such as rules to control commit messages or the author's email address.

You can learn more about rulesets here: `https://docs.github.com/en/repositories/ configuring-branches-and-merges-in-your-repository/managing-rulesets/ about-rulesets`.

Building different versions using a matrix

In this recipe, we are going to build and test our software for different versions, in our case, of the NodeJS environment.

Getting ready

Make sure you have cloned the repository from the previous recipe. Create a new branch to modify the workflow:

```
$ git switch -c build-matrix
```

Open the `.github/workflows/ci.yml` file in an editor.

How to do it...

1. Add the following code to the workflow file:

    ```
    strategy:
      matrix:
        node-version: ["21.x", "20.x"]
    ```

    Adjust the versions if needed.

2. In the `actions/setup-node` action, set the node version to the corresponding value from the matrix context:

    ```
    - uses: actions/setup-node@v4
      with:
        node-version: ${{ matrix.node-version }}
        check-latest: true
    ```

3. Commit and push your changes and create a pull request:

    ```
    $ git add .
    $ git commit
    $ git push -u origin build-matrix
    $ gh pr create --fill
    ```

4. Check the output of the workflow. It will run a separate job for each entry in the matrix array (see *Figure 6.3*):

Summary

Jobs

✅ build (21.x)

✅ build (20.x)

Figure 6.3 – The matrix runs a different job for each entry

5. Wait until the workflow has completed. Merge your pull request and clean up your repository:

```
$ gh pr merge -m
```

How it works...

The matrix is a convenient way to use the same workflow jobs with different combinations. It can contain one or multiple arrays that can contain many values. The matrix will run all combinations of all values in all arrays. You can think of the matrix as nested for loops. A good example is running and testing different versions on different platforms:

```
jobs:
  example_matrix:
    strategy:
      matrix:
        os: [ubuntu-22.04, ubuntu-20.04]
        version: [10, 12, 14]
    runs-on: ${{ matrix.os }}
    steps:
      - uses: actions/setup-node@v3
        with:
          node-version: ${{ matrix.version }}
```

This way, you can reuse the same workflow logic to test many different combinations of values at the same time.

There's more...

The matrix has some additional features. You can set fail-fast to indicate if it will cancel the workflow if one job in the matrix fails or if it should continue. You can define the number of parallel jobs with max-parallel, and you can include and exclude values for certain elements. Here is a more complex example:

```
jobs:
  test:
    runs-on: ubuntu-latest
```

```
continue-on-error: ${{ matrix.experimental }}
strategy:
  fail-fast: true
  max-parallel: 2
  matrix:
    version: [5, 6, 7, 8]
    experimental: [false]
    include:
      - version: 9
        experimental: true
```

To learn more about the matrix strategy, see `https://docs.github.com/en/actions/using-jobs/using-a-matrix-for-your-jobs`.

Informing the user on details of your build and test results

In this recipe, we are going to decorate pull requests and workflow summaries with details of our test results. We'll also add badges to the README file to indicate the quality of a branch or release.

Getting ready

Create a new branch to do the modification:

```
$ git switch -c add-badges
```

How to do it...

You can download badges for workflows using the following URL:

```
https://github.com/OWNER/REPO/actions/workflows/FILE.yml/badge.svg
```

You can also filter by branch or event by adding query parameters (for example, `?branch=main` or `?event=push`). We want a badge for the `main` branch, so add the following image to the markdown of your README:

```
![main](https://github.com/OWNER/package-recipe/actions/workflows/ci.yml/badge.svg?branch=main)
```

The badge will use the name in the workflow file and look like *Figure 6.4* in the preview:

Figure 6.4 – Badge for the CI workflow

But we want to go one step further and add a code coverage badge to the README, annotate pull requests with the code coverage output, and add a summary to the workflow.

1. To decorate the pull requests, the workflow needs write permissions. And, as we build multiple versions with the matrix, we have to pick one that we use for the creation of the badge. Add the following code to the `build` job:

    ```
    jobs:
      build:
        permissions:
            pull-requests: write
        env:
          MAIN_VERSION: "21.x"
    ```

2. In the workflow file, add the following code right after `run: npm test`:

    ```
    - name: Prepare coverage report in markdown
      uses: fingerprintjs/action-coverage-report-md@v1.0.6
      id: coverage
      with:
        textReportPath: coverage/coverage.txt
    ```

    This uses the `fingerprintjs/action-coverage-report-md` action to create a report in markdown that we'll write to the job summary and the pull request.

3. Next, we use the `marocchino/sticky-pull-request-comment` action to write the markdown report to the pull request:

    ```
    - name: Add coverage comment to the PR
      uses: marocchino/sticky-pull-request-comment@v2.8.0
      with:
        message: ${{ steps.coverage.outputs.markdownReport }}
    ```

4. Also, write the output to `$GITHUB_STEP_SUMMARY`. As we run the job in a matrix, add a heading indicating the version number:

    ```
    - name: Add coverage report to the job summary
      run: |
          echo "## Code Coverage v${{ matrix.node-version }}" >>
      "$GITHUB_STEP_SUMMARY"
          echo "${{ steps.coverage.outputs.markdownReport }}" >>
      "$GITHUB_STEP_SUMMARY"
    ```

5. Commit and push your changes and create a pull request:

```
$ git add .
$ git commit
$ git push -u origin add-badges
$ gh pr create -fill
```

This will trigger the workflow with the pull request trigger, and you can inspect in the workflow run the job summary, which should look like *Figure 6.5*:

build (21.x) summary   ...

Code Coverage v21.x

St	File	% Stmts	% Branch	% Funcs	% Lines	Uncovered Line #s
⬤	All files	100	100	100	100	
⬤	index.js	100	100	100	100	

Job summary generated at run-time

Figure 6.5 – Displaying code coverage results in the workflow summary

The summary is also added as a comment to the pull request (see *Figure 6.6*):

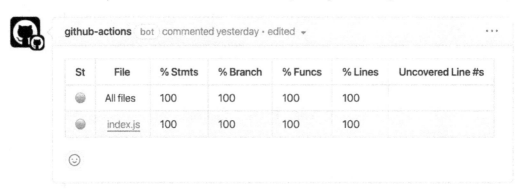

| github-actions | bot | commented yesterday · edited ▾ | | | | ... |

St	File	% Stmts	% Branch	% Funcs	% Lines	Uncovered Line #s
⬤	All files	100	100	100	100	
⬤	index.js	100	100	100	100	

☺

Figure 6.6 – Adding code coverage results as a pull request comment

6. The badge for code coverage is automatically created by the npm package during test. But in order to add the badge (that is, to the README), we need a place to host it. We'll use **GitHub Pages** for this.

Add the following code to the end of the `build` job to upload the artifacts to the workflow. Note that this is not the normal `actions/upload-artifact` action – it is a special action for GitHub Pages:

```
- name: Upload page artifacts
  if: ${{ matrix.node-version == env.MAIN_VERSION }}
  uses: actions/upload-pages-artifact@v3
  with:
    path: coverage
```

As we can only upload one artifact with the same name, we only run this step if the node version of the matrix is our main version.

7. Add a new `deploy` job after `build` that only runs on pushes to `main` and depends on the `build` job:

```
deploy:
  if: ${{ github.ref == 'refs/heads/main' }}
  needs: build
  runs-on: ubuntu-latest
```

8. Create a concurrency group called `pages` so that we only deploy one version at a time to the pages environment. But don't cancel deployments that are in progress to deploy all versions:

```
concurrency:
  group: "pages"
  cancel-in-progress: false
```

9. The job needs the following permissions:

```
permissions:
  contents: read
  pages: write
  id-token: write
```

10. The job deploys to the `github-pages` environment and uses the URL from the `actions/deploy-pages` action to display it in the workflow. The URL will point to the root of the package. As the HTML report with our `index.html` file is in the `lcov-report` folder, we add this to the URL:

```
environment:
  name: github-pages
  url: "${{ steps.deployment.outputs.page_url }}lcov-report"
```

11. The job has only two simple steps – `configure-pages` and `deploy-pages`:

```
steps:
  - name: Setup Pages
```

```
         uses: actions/configure-pages@v4
       - name: Deploy to GitHub Pages
         id: deployment
         uses: actions/deploy-pages@v4
```

12. Commit and push your changes and merge the pull request after the workflow has completed:

```
$ git add .
$ git commit
$ git push
$ gh merge -m
```

13. After the pull request is merged, a new workflow run will be triggered by the push to main. Once it is completed, you can see the URL to the website that has the report (see *Figure 6.7*):

Figure 6.7 – Inspecting the Pages website

14. The website created by GitHub contains the badge in the root of the directory. The URL looks like this: https://{OWNER}.github.io/package-recipe/badge.svg. The URL of the badge in markdown would look like this:

```
[![Coverage](https://wulfland.github.io/package-recipe/
badge.svg)]
```

But as we want to be directed to our lcov-report folder in the website when clicking on the badge, we'll add a link around the image. Add the following markdown to your README (replace wulfland with your GitHub username):

```
[![Coverage](https://wulfland.github.io/package-recipe/badge.
svg)](https://wulfland.github.io/package-recipe/lcov-report)
```

The badge will look like *Figure 6.8*, and when you click on it, you'll be redirected to the coverage report:

Figure 6.8 – Adding a badge for test coverage

How it works...

Let's understand how the code works.

Adding a workflow status badge

You can display a status badge for workflows in your repository to indicate the status of the workflows.

The URL of the badge is the following, and it will use the name of the workflow and show the status of the last workflow run:

```
https://github.com/OWNER/REPOSITORY/actions/workflows/WORKFLOW-FILE/
badge.svg
```

Typically, you only want to display the status for a specific branch or event. You can do so by providing the corresponding parameters (that is, `?branch=main` or `?event=push`). This way, you can display multiple badges for different versions of your software. See `https://docs.github.com/en/actions/monitoring-and-troubleshooting-workflows/adding-a-workflow-status-badge` for more information.

GitHub Pages

GitHub Pages is a static site hosting service from GitHub that takes static files from a repository and publishes them as a website. The site is hosted on GitHub's `github.io` domain, or you can use your own custom domain.

Unless you're using a custom domain, project sites are available at the following URLs:

- `http(s)://<username>.github.io/<repository>` or
- `http(s)://<organization>.github.io/<repository>`.

Pages can be deployed directly from a branch and, optionally, prepossessed with Jekyll (`https://github.com/jekyll/jekyll`). This way, you can easily render markdown files as a website – for example, to host a blog.

You can find an example under `https://wulfland.github.io/AccelerateDevOps/` that renders the content of the `https://github.com/wulfland/AccelerateDevOps/tree/main/docs` folder.

In addition to deploying from a branch, you can also deploy pages using your own workflows. This is what we did in this recipe. The feature is still in beta at the time of writing – but in my opinion, it can already be used in production.

See `https://docs.github.com/en/pages/getting-started-with-github-pages/about-github-pages` to learn more about GitHub Pages.

Concurrency groups

By default, GitHub Actions allows multiple jobs within the same workflow and multiple workflow runs within the same repository to run concurrently – meaning that multiple steps can run at the same time.

GitHub Actions also allows you to control the concurrency of workflow runs so that you can ensure that only one run, one job, or one step runs at a time in a specific context. This can be useful for controlling situations when running multiple steps at the same time could cause conflicts or consume more action minutes than expected.

Concurrency groups can have a static name, like we did in our recipe. But you can also use context expressions to group certain contexts – such as branches – together in one group:

```
concurrency:
  group: ${{ github.workflow }}-${{ github.ref }}
  cancel-in-progress: true
```

The `cancel-in-progress` parameter can be used to cancel a job if a newer version is available and run that instead. This can help save resources. If it is set to `false`, the workflow will wait until the previous version has completed and then run the next job in the concurrency queue. This way, only one job at a time gets executed for a given group.

You can learn more about concurrency groups here: `https://docs.github.com/en/actions/using-jobs/using-concurrency`.

There's more...

Validating code automatically and bringing detailed information to where it is valuable for the developers is very important. In addition to test results, code coverage, and linters, you can also integrate solutions such as SonarQube or SonarCloud. They provide a great GitHub integration and are free for open source. Just log in with your GitHub credentials, and you can configure a new project. The project allows you to create additional badges that you can add to your README (see *Figure 6.9*):

Code scanning

Automatically detect common vulnerabilities and coding errors.

Figure 6.9 – Using SonarCloud badges in your GitHub README

Sonar badges provide a great integration into your pull request and can be used as an additional status check in rulesets and branch protection rules. The quality gates are reported in the pull request (see *Figure 6.10*):

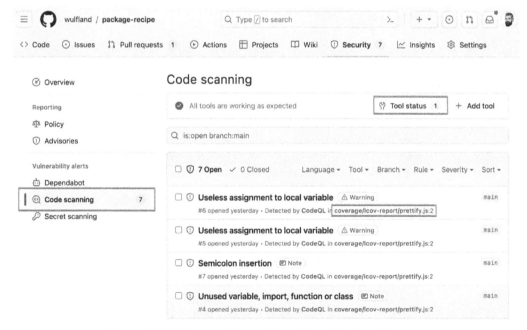

Figure 6.10 – Pull request integration of SonarCloud quality gates

To learn more about GitHub integration for SonarCloud, see the documentation at `https://docs.sonarsource.com/sonarcloud/getting-started/github/`.

Finding security vulnerabilities with CodeQL

This is a short recipe that we will use to add a **CodeQL** analysis to our existing CI build. CodeQL is the code analysis engine from GitHub, and it is free for public repositories.

Getting ready

Create a new branch in the `package-recipe` repository:

```
$ git switch -c add-codeql
```

How to do it...

1. Open the `.github/workflows/ci.yml` file and grant the `build` job permissions to write security events:

    ```
    build:
      permissions:
        pull-requests: write
        security-events: write
    ```

2. Add an `init` action (`github/codeql-action/init`) to the job. For languages that must be compiled, this has to go before the build process. As JavaScript is a static language, you can add it to the end of the job. Set the language to `javascript-typescript` and select the `security-and-quality` query suite:

    ```
    - name: Initialize CodeQL
      uses: github/codeql-action/init@v3
      with:
        languages: 'javascript-typescript'
        queries: security-and-quality
    ```

3. Performing the actual **analysis** is executed after compilation. Just add the following code to the end of the job:

    ```
    - name: Perform CodeQL Analysis
      uses: github/codeql-action/analyze@v3
      with:
        category: "/language:javascript-typescript"
    ```

4. Commit your code, create a pull request, and merge the changes when all the checks have passed:

```
$ git add .
$ git commit
$ gh pr create --fill
$ gh pr merge -m --auto
```

How it works...

CodeQL is the code analysis engine from GitHub to automate security and quality checks. It is free in public repositories, and it is available for enterprises that purchase the GitHub Advanced Security license.

You can analyze the following languages:

- C/C++

- C#

- Go

- Java

- Kotlin (in beta at the time of writing)

- JavaScript/TypeScript

- Python

- Ruby

- Swift (in beta at the time of writing)

There are three ways to analyze your code:

- **Default**: The default setup for CodeQL will automatically detect the languages in your repository and pick the supported ones to analyze. It will apply the default query suites and triggers for the scans. In the repository, under **Settings | Code security and analysis**, you can select **Default** if the languages in your repository are supported by that feature (see *Figure 6.11*):

Code scanning

Automatically detect common vulnerabilities and coding errors.

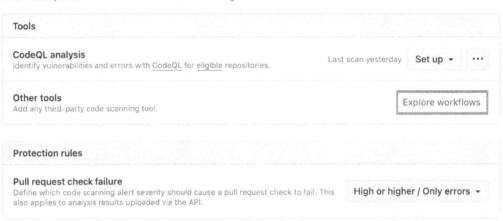

Figure 6.11 – Configuring code scanning with Default or Advanced mode

- **Advanced**: Use a custom workflow and add the CodeQL analysis as we did in the recipe. If you click on **Advanced** in **Settings** (*Figure 6.11*), this will generate a customizable workflow template for you. It will use a matrix strategy with `fail-fast: false` to analyze all languages detected in your repository.

- **CLI**: Run the CodeQL CLI directly in an external CI system and upload the results to GitHub.

The results of the analysis of all tools can be accessed under **Security | Code scanning** (see *Figure 6.12*):

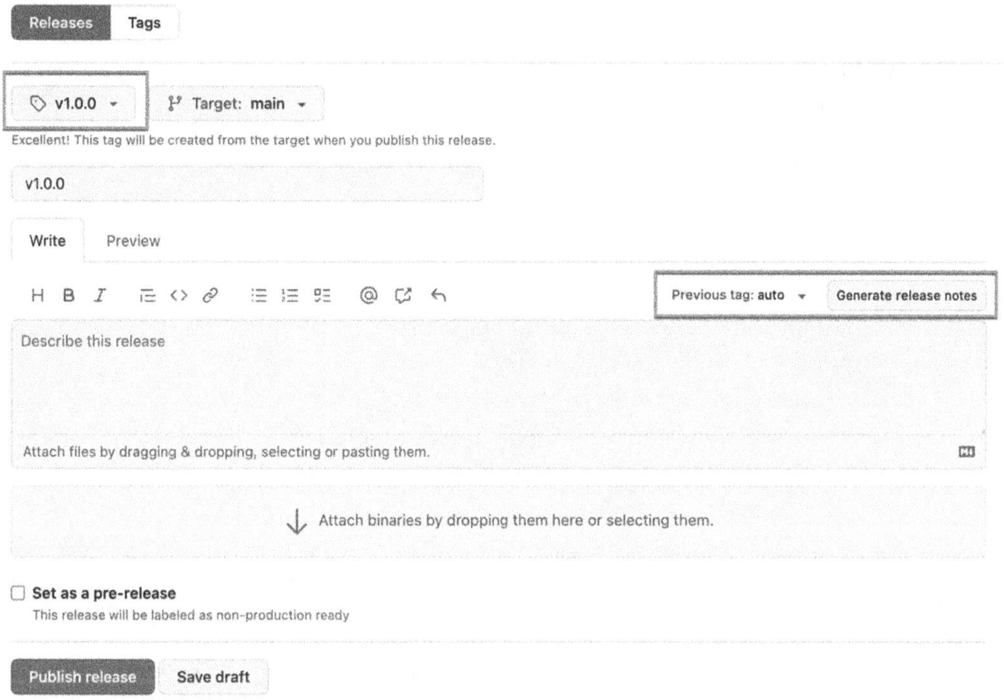

Figure 6.12 – Viewing code analysis results in a repository

You can access the status of all tools that report results. Note that in our repository, the analysis also analyses code generated by our test packages.

To learn more about CodeQL, visit `https://docs.github.com/en/code-security/code-scanning/introduction-to-code-scanning/about-code-scanning-with-codeql`.

There's more...

GitHub supports all kinds of other tools – among others, **Checkmarx**, **Snyk**, **Microsoft Defender for DevOps**, and **ESLint**. It supports many commercial ones but also many that are free or open source.

If you click on **Explore workflows** in your repository under **Settings | Code security and analysis** (see *Figure 6.13*), then you will be redirected to a new workflow page filtered by the `security` category and `code scanning` query (`https://github.com/OWNER/REPO/actions/new?category=security&query=code+scanning`):

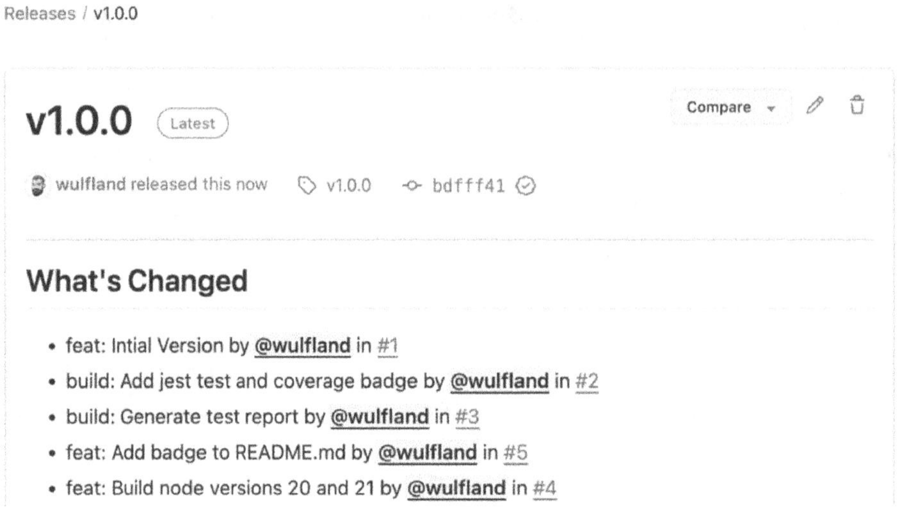

Releases / v1.0.0

v1.0.0   (Latest)

Compare ▾   ✎   🗑

🧑 wulfland released this now   🏷 v1.0.0   🔗 bdfff41 ⊘

What's Changed

- feat: Intial Version by **@wulfland** in #1
- build: Add jest test and coverage badge by **@wulfland** in #2
- build: Generate test report by **@wulfland** in #3
- feat: Add badge to README.md by **@wulfland** in #5
- feat: Build node versions 20 and 21 by **@wulfland** in #4

Figure 6.13 – Adding third-party scanning tools to your repository

You can explore all partners that have published a template workflow to the marketplace.

If you have other tools, you can still integrate them as long as they support **Static Analysis Results Interchange Format** (**SARIF**; see `https://sarifweb.azurewebsites.net/`) – the approved **Organization for the Advancement of Structured Information Standards** (**OASIS**) standard for static code analysis – as an output format.

To give you an example, you could use **Checkov**, a static code analysis tool for **infrastructure as code** (**IaC**) supporting Terraform, Terraform plan, CloudFormation, AWS **Serverless Application Model** (**SAM**), Kubernetes, Helm charts, Kustomize, Dockerfiles, Serverless, Bicep, OpenAPI or ARM templates. Use the action `bridgecrewio/checkov-action` action to run your analysis and then upload the results to GitHub:

```
- name: Checkov GitHub Action
  uses: bridgecrewio/checkov-action@v12
  with:
    output_format: sarif

- name: Upload SARIF file
  uses: github/codeql-action/upload-sarif@v3
  with:
    sarif_file: results.sarif
  if: always()
```

This way, you can easily integrate all code scanning tools available on the market, as long as they support SARIF.

Creating a release and publishing the package

In this recipe, we are going to create a workflow that will publish our package whenever a release is created.

Getting ready

Create a new branch:

```
$ git switch -c add-release-workflow
```

How to do it...

1. Create a new `.github/workflows/release.yml` workflow file. Name the workflow `Release` and trigger it on the creation of GitHub releases:

    ```
    name: Release

    on:
      release:
        types: [created]
    ```

2. Add a `publish` job and give it read permission for the repository and write for packages:

    ```
    jobs:
      publish:
        runs-on: ubuntu-latest
        permissions:
          packages: write
          contents: read
    ```

3. Check out the code and configure the NodeJS environment with the correct version. Also, set the registry to use GitHub:

    ```
    steps:
      - uses: actions/checkout@v4

      - uses: actions/setup-node@v4
        with:
          node-version: 21.x
          registry-url: https://npm.pkg.github.com/
    ```

4. Build and test the application and publish it to the registry using a GitHub token:

    ```
    - name: Install dependencies
      run: npm install
    ```

```
- name: Run tests
  run: npm test

- run: npm publish
  env:
    NODE_AUTH_TOKEN: ${{secrets.GITHUB_TOKEN}}
```

5. Commit your changes, create a pull request, and merge the changes once the checks have passed:

    ```
    $ git add .
    $ git commit
    $ gh pr create --fill
    $ gh pr merge -m --auto
    ```

6. Once your pull request has been merged, go to **Releases** on the **Code** tab of your repository and click **Draft a new Release**.

7. Click **Choose a tag**, enter v1.0.0, and click **Create new tag**. Instead of adding the body of the release in markdown, just click **Generate release notes** (see *Figure 6.14*):

Contributors

wulfland

▼ Assets   2

🗂️ Source code (zip)	
	8 minutes ago
🗂️ Source code (tar.gz)	
	8 minutes ago

☺

Figure 6.14 – Drafting a new release for a tag and generating release notes

This will generate details for your release from your pull requests. Click **Publish release** to create the release. The release should look like *Figure 6.15*:

Figure 6.15 – Details of a GitHub release

8.   It should also contain the source code as `.zip` and `.tar` archives per default (see *Figure 6.16*):

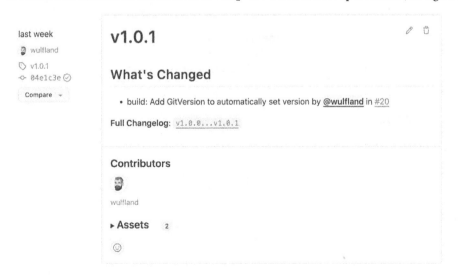

Figure 6.16 – Assets of a release contain the source code as .zip and .tar archives

9. The creation of the release will trigger the workflow, and it will publish your package with version 1.0.0 to your repository (note that this comes from your package.json file and not the tag!).

The package looks like *Figure 6.17*:

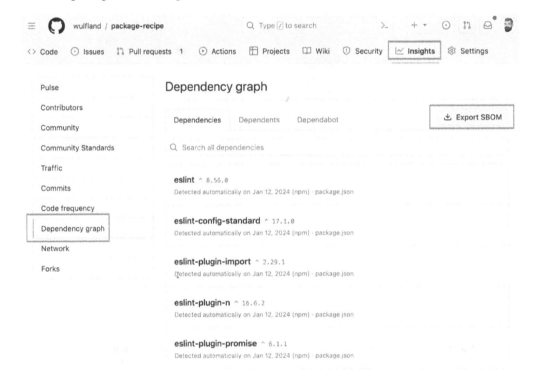

Figure 6.17 – The package in the repository

It contains installation instructions, information on your repository, and the README with your badges.

How it works...

Now, we'll see how it works.

Semantic versioning

Packages are typically created using **semantic versioning**, a formal convention for specifying version numbers for software. It consists of different parts with different meanings. Examples of semantic version numbers are 1.0.0 or 1.5.99-beta. The format is as follows:

```
<major>.<minor>.<patch>-<pre>
```

- **Major version**: A numeric identifier that gets increased if the version is not backward compatible and has breaking changes. An update to a new major version must be handled with caution! A major version of zero is for the initial development.

- **Minor version**: A numeric identifier that gets increased if new features are added, but the version is backward compatible with the previous version and can be updated without breaking anything if you need the new functionality.

- **Patch**: A numeric identifier that gets increased if you release bug fixes that are backward compatible. New patches should always be installed.

- **Pre-version**: A text identifier that is appended using a hyphen. The identifier must only use ASCII alphanumeric characters and hyphens ([0-9A-Za-z-]). The longer the text, the smaller the pre-version (meaning -alpha < -beta < -rc). A pre-release version is always smaller than a normal version (1.0.0-alpha < 1.0.0).

See https://semver.org/ for the complete specification.

In our case, the semantic version for the package was set in the package.json file. In the next recipe, we will use the release process to automatically set the version number of the package to the release so that you do not have to do this manually.

Releases

You can create a release to package software, release notes, and binary files for other people to download.

GitHub releases are deployable software versions you can package and make available for a wider audience to download and use. They can contain release notes and other binary files to download.

Releases are based on **Git tags**, which mark a specific point in your repository's history.

In the next recipes, we will use GitHub Releases together with tags to automatically version our package and to attach an SBOM in an automated way. To learn more about GitHub Releases, see https://docs.github.com/en/repositories/releasing-projects-on-github.

There's more...

Releasing and versioning of software depends a lot on your workflow – especially how you work with `git` branches and tags. In our example, I assume that you start a release process by directly creating a release. You could also use the push of a tag:

```
on:
  push:
    tags:
      - v*.**
```

The workflow would be triggered by any push of a tag that starts with v, and you could use it to automatically create a release:

```
gh release create ${{ github.ref_name }} --generate-notes
```

You could also do this on pushes to a `release` branch:

```
on:
  push:
    branches:
      - release/*
```

Versioning your packages

If you now created a new release, the workflow would fail as it would try to publish version 1.0.0 again to the package registry. You would manually have to set the version number in the `package.json` file. In this recipe, we will use **GitVersion** to automate this process.

Getting ready

Switch to a new branch:

```
$ git switch -c add-gitversion
```

How to do it...

1. For **GitVersion** to automatically determine the version of your `git` workflow, you have to download all references and not just the HEAD branch. We do this by adding the `fetch-depth` parameter to the checkout action and setting it to 0:

    ```
    - uses: actions/checkout@v4
      with:
        fetch-depth: 0
    ```

2. Set up `GitVersion` in a specific version:

    ```
    - name: Install GitVersion
      uses: gittools/actions/gitversion/setup@v0.10.2
      with:
        versionSpec: '5.x'
    ```

3. Execute `GitVersion` to determine the version number:

    ```
    - name: Determine Version
      uses: gittools/actions/gitversion/execute@v0.10.2
    ```

4. Change the version of the npm package:

    ```
    - name: 'Change NPM version'
      uses: reedyuk/npm-version@1.2.2
      with:
        version: $GITVERSION_SEMVER
    ```

5. Commit your changes, create a pull request, and merge it when all checks have passed:

    ```
    $ git add .
    $ git commit -m '(build): Add GitVersion to automatically set version'
    $ gh pr create --fill
    $ gh pr merge -m --auto
    $ gh pr checks
    ```

6. Wait until all checks have passed. Then, create the release:

    ```
    $ gh release create v1.0.1 --generate-notes
    ```

    You will have a new release like *Figure 6.18* – this time created from the CLI:

Figure 6.18 – Creating a release from the CLI that will trigger the workflow
to publish a new package with the same version as the release

The release will trigger the workflow, and a new package with version 1.0.1 will be released.

How it works...

GitVersion (see https://gitversion.net/docs/) is a tool that automatically generates a semantic version number based on your Git history. GitVersion runs with a default configuration that works with **GitHub flow** (https://docs.github.com/en/get-started/using-github/github-flow) and **Git flow** (https://nvie.com/posts/a-successful-git-branching-model/). You can run GitVersion init to launch a wizard that guides you through creating a config file (GitVersion.yml). In our example, we use the Continuous Delivery mode – meaning we explicitly create a version using a tag. But there are also other modes such as Continuous Deployment mode (creating a version from every commit to specific branches) or Mainline mode.

The gittools/actions/gitversion/execute action will execute GitVersion and save the result in the $GITVERSION_SEMVER environment variable. You can also access individual parts of the version and configuration using the output of the action:

```
- name: Determine Version
  id: gitversion
  uses: gittools/actions/gitversion/execute@v0

- name: Display GitVersion outputs (step output)
```

```
run: |
  echo "Major: ${{ steps.gitversion.outputs.major }}"
  echo "Minor: ${{ steps.gitversion.outputs.minor }}"
  echo "Patch: ${{ steps.gitversion.outputs.patch }}"
```

There's more...

You can also use **Conventional Commits** (see https://www.conventionalcommits.org) to automatically determine semantic versions out of commit messages. Conventional Commits is a specification that provides a set of rules for creating an explicit commit history by describing features, fixes, and breaking changes made in commit messages. You can use Conventional Commits to create release notes – and you can use it with GitVersion to automatically determine if a new version is a patch, minor, or major number. This is done in the `GitVersion.yml` configuration:

```
mode: Mainline
major-version-bump-message:
"^(build|chore|ci|docs|feat|fix|perf|refactor|revert|style|test)(\\
([\\w\\s-]*\\))?(!:|:.*\\n\\n((.+\\n)+\\n)?BREAKING CHANGE:\\s.+)"
minor-version-bump-message: "^(feat)(\\([\\w\\s-]*\\))?:"
patch-version-bump-message:
"^(build|chore|ci|docs|fix|perf|refactor|revert|style|test)(\\([\\w\\
s-]*\\))?:"
```

In the CI build, you can then add a job that determines the version out of the Conventional Commit messages and creates a new release:

```
publish:
  if: ${{ github.ref == 'refs/heads/main' }}
  needs: build
  runs-on: ubuntu-latest
  permissions:
    contents: write

  steps:
    - uses: actions/checkout@v4
      with:
        fetch-depth: 0

    - name: Install GitVersion
      uses: gittools/actions/gitversion/setup@v0.10.2
      with:
        versionSpec: '5.x'

    - name: Determine Version
      uses: gittools/actions/gitversion/execute@v0.10.2
```

```
      with:
        useConfigFile: true

    - name: Create a new release
      env:
        GH_TOKEN: ${{ github.token }}
      run: |
        gh release create ${{ env.GITVERSION_SEMVER }}
  --generate-notes
```

The two workflows together will then completely automate the creation of new versions after every merge to `main`.

Generating and using SBOMs

An SBOM (see `https://www.cisa.gov/sbom`) declares the nested inventory of components that make up the software. The United States government is required to obtain an SBOM for any product they purchase by the Cyber Supply Chain Management and Transparency Act of 2014.

You can manually export an SBOM in GitHub under **Insights | Dependency graph | Export SBOM** (see *Figure 6.19*):

Figure 6.19 – Manually exporting an SBOM in a repository

The SBOM is a JSON file following the **Software Package Data Exchange** (**SPDX**) standard.

In this recipe, we will automate the process of generating an SBOM from the current dependencies of the repository and attach it to the release as an additional attachment.

Getting ready

Switch to a new branch:

```
$ git switch -c upload-sbom
```

How to do it...

1. Edit the `.github/workflows/release.yml` file. Modify the permission for the `publish` job to allow write access to permissions:

    ```
    jobs:
      publish:
        runs-on: ubuntu-latest
        permissions:
          packages: write
          contents: write
    ```

2. After the `Publish package` step, add the following step, which calls the API using the GitHub CLI and downloads the SBOM:

    ```
    - name: Generate SBoM
      env:
        GH_TOKEN: ${{ github.token }}
      run: |
        gh api \
          -H "Accept: application/vnd.github+json" \
          -H "X-GitHub-Api-Version: 2022-11-28" \
          /repos/wulfland/package-recipe/dependency-graph/sbom >
    sbom.json
    ```

3. Add the `svenstaro/upload-release-action` action and upload the SBOM as an attachment to the release:

    ```
    - name: Upload SBoM to release
      uses: svenstaro/upload-release-action@v2
      with:
        file: sbom.json
        asset_name: SBoM
        tag: ${{ github.ref }}
        overwrite: true
    ```

4. Commit the workflow, create a pull request, and merge it back once the checks have passed:

```
$ git add .
$ git commit
$ gh pr create --fill
$ gh pr merge -m --auto
$ gh pr checks
```

5. Wait until all checks have passed. Then, create a release:

```
$ gh release create v1.0.5 --generate-notes
```

6. The workflow will create a release and attach the SBOM as an asset (see *Figure 6.20*):

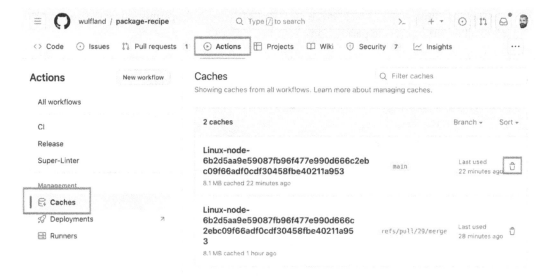

Figure 6.20 – Automatically attached SBOM as an asset in the release

You can download and inspect the JSON file from the release.

How it works...

There is always the need to store output – sometimes binary and sometimes text files, as in this case – when releasing software.

You can upload artifacts to the GitHub workflow using the `https://github.com/actions/upload-artifact` action:

```
- uses: actions/upload-artifact@v4
  with:
    name: my-artifact
    path: path/**/[abc]rtifac?/*
```

We did this with the test badges to publish them to GitHub Pages. You can then download the artifacts from the workflow summary page or in subsequent jobs using the `https://github.com/actions/download-artifact` action:

```
- uses: actions/download-artifact@v4
  with:
    name: my-artifact
```

But for end users, it is more convenient to store everything in a release. This way, you bundle all build output with an exact version of your source code. You can manage release assets using the API (see `https://docs.github.com/en/rest/releases/assets`):

```
gh api \
  --method POST \
  -H "Accept: application/vnd.github+json" \
  -H "X-GitHub-Api-Version: 2022-11-28" \
  --hostname HOSTNAME \
  /repos/OWNER/REPO/releases/ID/assets?name=example.zip \
  -f '@example.zip'
```

But there are many actions available that help with that. We use the `svenstaro/upload-release-action` action – but there are many others. For example, **GitReleaseManager** from **GitTools** (see `https://github.com/GitTools/actions/blob/main/docs/examples/github/gitreleasemanager/index.md`) has an entire suite of actions to interact with releases.

There's more...

There are different common formats for SBOM:

- **SPDX**: SPDX is an open standard for SBOM with origins in the Linux Foundation. Its origin was license compliance, but it also contains copyrights, security references, and other metadata. SPDX was recently approved as an ISO/IEC standard (*ISO/IEC 5962:2021*), and it fulfills the **National Telecommunications and Information Administration's** (**NTIA's**) *The Minimum Elements For a Software Bill of Materials*.

- **CycloneDX (CDX)**: CDX is a lightweight open source format with origins in the **Open Worldwide Application Security Project** (**OWASP**) community. It is optimized for integrating SBOM generation into a release pipeline.

- **Software Identification (SWID) tags**: SWID is an ISO/IEC industry standard (*ISO/IEC 19770-2*) used by various commercial software publishers. It supports automation of software inventory, assessment of software vulnerabilities on machines, detection of missing patches, targeting of configuration checklist assessments, software integrity checking, installation and execution whitelists/blacklists, and other security and operational use cases. It is a good format for doing the inventory of the software installed on your build machines.

There are different tools and use cases for each format. **SPDX** is used by **GitHub**, **FOSSology**, and **syft**. You can use the **Anchore SBOM Action** (see `https://github.com/marketplace/actions/anchore-sbom-action`) to generate an SPDX SBOM for a Docker or **Open Container Initiative** (**OCI**) container:

```
- name: Anchore SBOM Action
    uses: anchore/sbom-action@v0.6.0
    with:
      path: .
      image: ${{ env.REGISTRY }}/${{ env.IMAGE_NAME }}
      registry-username: ${{ github.actor }}
      registry-password: ${{ secrets.GITHUB_TOKEN }}
```

The SBOM is being uploaded as a workflow artifact.

CDX (`https://cyclonedx.org/`) is more focused on application security. There are versions for **NodeJS**, **.NET**, **Python**, **PHP**, and **Go** in the marketplace – but many more languages are supported using the CLI or other package managers (**Java**, **Maven**, **Conan**, and many more). The usage is simple. Here is an example of the action for .NET:

```
- name: CycloneDX .NET Generate SBOM
    uses: CycloneDX/gh-dotnet-generate-sbom@v1.0.1
    with:
      path: ./CycloneDX.sln
      github-bearer-token: ${{ secrets.GITHUB_TOKEN }}
```

The SBOM does not get uploaded automatically unlike the Anchore action – you would have to do that manually:

```
- name: Upload a Build Artifact
    uses: actions/upload-artifact@v2.3.1
    with:
      path: bom.xml
```

CDX is also used in **OWASP Dependency-Track** (see `https://github.com/DependencyTrack/dependency-track`) – a component analysis platform that you can run as a container or in Kubernetes. You can upload the SBOM directly into your *Dependency-Track* instance:

```
- uses: DependencyTrack/gh-upload-sbom@v1.0.0
  with:
    serverhostname: 'your-instance.org'
    apikey: ${{ secrets.DEPENDENCYTRACK_APIKEY }}
    projectname: 'Your Project Name'
    projectversion: 'main'
```

SWID tags are more used in **software asset management (SAM)** solutions such as Snow (`https://www.snowsoftware.com/`), **Microsoft System Center**, or **ServiceNow IT Operations Management (ITOM)**. **CDX** and **SPDX** can use SWID tags if they are present.

If you want to learn more about SBOM, see `https://www.ntia.gov/sbom`.

Using caching in workflows

In this recipe, we will use caching to optimize the speed of workflows.

Getting ready

Switch to a new branch:

```
$ git switch -c cache-npm-packages
```

How to do it...

1. Edit the `.github/workflows/ci.yml` file. **After** the `setup-node` action, add the following script to get the npm cache directory for the correct npm version and store it as an output variable:

    ```
    - name: Get npm cache directory
      id: npm-cache-dir
      run: echo "dir=$(npm config get cache)" >> "${GITHUB_OUTPUT}"
    ```

2. Add the actual cache step right after that. Also, give it a name and create a key from the hash of the `package-lock.json` file:

    ```
    - uses: actions/cache@v3
      id: npm-cache
      with:
        path: ${{ steps.npm-cache-dir.outputs.dir }}
        key: ${{ runner.os }}-node-${{ hashFiles('**/package-lock.
    ```

```
json') }}
    restore-keys: |
        ${{ runner.os }}-node-
```

3. The next step will just list the dependencies if they are added to the cache:

```
- name: List the state of node modules
  if: ${{ steps.npm-cache.outputs.cache-hit != 'true' }}
  continue-on-error: true
  run: npm list
```

4. Commit your changes and create a pull request:

```
$ git add.
$ git commit
$ gh pr create --fill
```

5. Open the CI workflow run in your pull request and inspect the output of the steps we added. Rerun both jobs and note how the action restores the packages from the cache and how `cache-hit` prevents the next step from executing (see *Figure 6.21*):

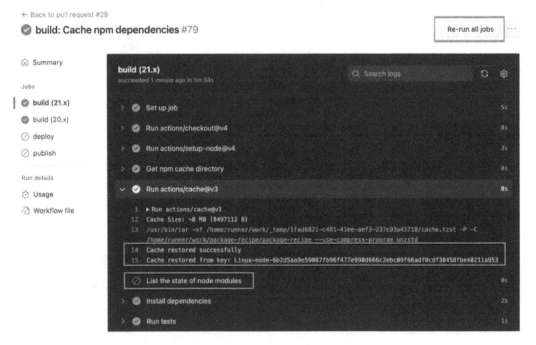

Figure 6.21 – The cache is successfully restored and cache-hit is true

6. Merge your pull request:

```
$ gh pr merge -s --auto
```

How it works...

It is important to only use the cache if you have performance problems. If you use the setup actions, you probably not going to see a big improvement. But it is important to know how caching works to have the tool available if you need it.

The cache action (`https://github.com/actions/cache`) stores information on GitHub-owned cloud storage and retrieves it in subsequent runs from that.

Let's assume you have a long-running operation – such as calculating prime numbers – and you use the output in a subsequent step. (I just use a sleep here to simulate a long-running task):

```
- name: Generate Prime Numbers
  run: |
    sleep 60
    echo "1 2 3..." > primes

- name: Use Prime Numbers
  run: cat primes
```

You can now add a cache step before that step and cache the folder primes:

```
- name: Cache Primes
  id: cache-primes
  uses: actions/cache@v3
  with:
    path: primes
    key: ${{ runner.os }}-primes
```

The key here is what has to be unique to determine if the cache has changed. In our example, we used the hash value of the `packe-lock.json` file. For prime numbers, we might have a different format for the operating system.

In the long-running operation, you can now check if the cache is valid and only execute it, when there is no hit for the current key:

```
- name: Generate Prime Numbers
  if: steps.cache-primes.outputs.cache-hit != 'true'
  run: |
    sleep 60
    echo "1 2 3..." > primes
```

This workflow would run the first time for the long-running operation and populate the file in the cache. If you run it a second time, it will load the file from the cache and omit the long-running step.

There's more…

A repository can store up to 10 GB of data in caches. Once that limit is reached, older files will be removed based on when they were last accessed. Caches that are not used for 1 week will also be cleaned up.

You can manage the cache under **Actions | Caches** (see *Figure 6.22*):

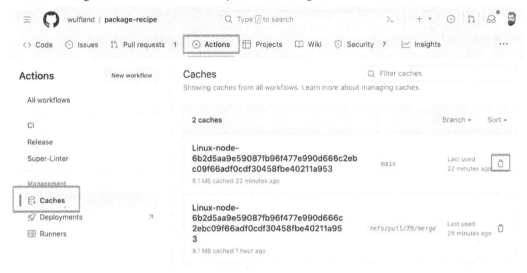

Figure 6.22 – Managing the cache

The cache action (`https://github.com/actions/cache`) has good documentation and examples for most programming languages. You can also refer to the documentation when you want to implement caching (see `https://docs.github.com/en/actions/using-workflows/caching-dependencies-to-speed-up-workflows`).

Release Your Software with GitHub Actions

In this final chapter, we will dive into using GitHub Actions for **continuous deployments** (**CDs**). We will create a container with a simple website that uses the package from *Chapter 6, Build and Validate Your Code*, and we will deploy it to Kubernetes in the cloud, securing access with **OpenID Connect** (**OIDC**). We will use environments to secure the deployment and concurrency groups to control the flow of multiple workflows.

We will use **Microsoft Azure Kubernetes Service** (**AKS**) in this chapter as the production environment, but I will point you to documentation to carry out the same recipes with other cloud providers such as **Google Kubernetes Engine** (**GKE**) on **Google Cloud Platform** (**GCP**) or **Elastic Container Services** (**ECSs**) on **Amazon Web Services** (**AWS**).

We'll cover the following recipes:

- Building and publishing a container
- Using OIDC to securely deploy to any cloud
- Environment approval checks
- Releasing the container application to **Azure Kubernetes Service** (**AKS**)
- Automating the update of your dependencies

Technical requirements

For this chapter, you will need **Docker**, **Node.js**, and the **GitHub CLI**, either on your local machine, or you can just use **GitHub Codespaces**. For the Microsoft Azure part, you will need an Azure account. If you don't have one, just create a free trial account here: `https://azure.microsoft.com/en-us/free`. You can use the Azure CLI locally or just use **Cloud Shell** in the **Azure portal**.

You will also need a GitHub **personal access token** (**PAT**) with read and write permission for GitHub packages (`https://docs.github.com/en/authentication/keeping-your-account-and-data-secure/managing-your-personal-access-tokens#creating-a-personal-access-token-classic`).

Building and publishing a container

In this recipe, we are going to containerize a simple web application and push it to a container registry.

Getting ready...

Open the repository (`https://github.com/wulfland/release-recipe`) and click on **Use this template** to create a new copy out of it (direct link: `https://github.com/new?template_name=release-recipe&template_owner=wulfland`). Create a new public repository in your personal account, name it `release-recipe`, and clone the repository.

Open the `package.json` and adjust the author and repository URL.

Under dependencies, adjust the owner and version of the package recipe to your package from *Chapter 6, Build and validate your code*:

```
"dependencies": {
  "@wulfland/package-recipe": "^2.0.5",
  "express": "^4.18.2"
}
```

Replace the owner in the file `.npmrc`:

```
@wulfland:registry=https://npm.pkg.github.com
```

In a terminal, run the following commands:

```
$ npm login --registry https://npm.pkg.github.com
```

Enter your GitHub username and the PAT token with access to packages. Run the following commands:

```
$ npm install
$ npm start

> release-recipe@1.0.0 start
> node src/index.js

Server running at http://localhost:3000
```

Now, you should have a Node.js application running on port 3000. Open a browser, navigate to http://localhost:3000, and validate that is displays Hello World!. Stop the server by exiting the process (*CTRL+C*).

How to do it...

1. Create a new file, Dockerfile, in the root of the repository. Inherit your image from the node image and pick the 21-bullseye version. Create a folder, copy the repository content in it, and make it the working folder:

    ```
    FROM node:21-bullseye
    RUN mkdir -p /app
    COPY . /app
    WORKDIR /app
    ```

2. Note that you have to run npm install before building the Docker image to avoid storing your credentials in the container. Rebuild the npm package in the container:

    ```
    RUN npm rebuild
    ```

3. Expose port 3000 of our express website and run npm start as the start command of the container:

    ```
    EXPOSE 3000
    CMD [ "npm", "start"]
    ```

4. Next, build your container image locally and run it:

    ```
    $ npm install
    $ docker build -t hello-world-recipe .
    $ docker run -it -p 3000:3000 hello-world-recipe
    ```

5. Verify that the website runs on port 3000 on your local machine again.

6. Create a new workflow .github/workflows/publish.yml. Run it on pull requests and pushes to the main branch:

    ```
    name: Publish Docker Image

    on:
      push:
        branches: [ main ]
      pull_request:
        branches: [ main ]
    ```

7. Set the registry name and image name as environment variables:

```
env:
  REGISTRY: 'ghcr.io'
  IMAGE_NAME: '${{ github.repository }}'
```

8. Add a job with permissions for GITHUB_TOKEN to write packages and read content:

```
jobs:
  build-and-push-image:
    runs-on: ubuntu-latest

    permissions:
      packages: write
      contents: read
```

9. Add the following steps. Check out the repository and log in to the Docker registry:

```
- name: Checkout repository
  uses: actions/checkout@v4

- name: Log in to the Container registry
  uses: docker/login-action@v3
  with:
    registry: ${{ env.REGISTRY }}
    username: ${{ github.actor }}
    password: ${{ secrets.GITHUB_TOKEN }}
```

10. Extract the metadata for the image from the registry. We use the long SHA as the tag for the container. Give the step an id to later access the output:

```
- name: Extract metadata (tags, labels) for Docker
  id: meta
  uses: docker/metadata-action@v5
  with:
    images: ${{ env.REGISTRY }}/${{ env.IMAGE_NAME }}
    tags: |
      type=sha,format=long
```

11. Set up Node.js to use the correct version and registry. Then, build and test your code. You have to set the NODE_AUTH_TOKEN environment variable to the GITHUB_TOKEN to authenticate to the package registry and receive the npm package from the package recipe from *Chapter 6*:

```
- uses: actions/setup-node@v4
  with:
    node-version: 21.x
```

```
      registry-url: https://npm.pkg.github.com/

    - name: Build and test
      env:
        NODE_AUTH_TOKEN: ${{ secrets.GITHUB_TOKEN }}
      run: |
        npm install
        npm run test
```

12. Now, we are ready to build and push the Docker image. Use the output of the `meta` step to set the tags and labels:

```
    - name: Build and push Docker image
      uses: docker/build-push-action@v5
      with:
        context: .
        push: ${{ github.event_name != 'pull_request' }}
        tags: ${{ steps.meta.outputs.tags }}
        labels: ${{ steps.meta.outputs.labels }}
```

13. Commit and push your changes. After the workflow run, you will find the Docker image on the **Code** tab of your repository on the right side under **Packages**.

How it works...

I use **Express** (`https://expressjs.com/`) as a simple web framework to run a website. The website displays the content from our package; we'll leverage this in the upcoming recipes to automatically keep our dependencies up to date. The code is easy to understand:

```
const express = require('express');
const greet = require('@wulfland/package-recipe/src/index')
const app = express();
const port = 3000;

app.get('/', (req, res) => {
  res.send(greet());
});

app.listen(port, () => {
  console.log(`Server running at http://localhost:${port}`);
});
```

This time, I have omitted all linting and testing, as we have covered this already in *Chapter 6*. We need to containerize this application to deploy it to the cloud later.

To deploy a container to the cloud, you have to store it in a container registry. We are using GitHub packages here, but using the cloud-specific registries works the same way. You just have to configure a PAT token and cannot use `GITHUB_TOKEN`, but this is the only difference.

There's more...

The Docker meta-action (`https://github.com/docker/metadata-action`) can be used to extract metadata from Git references and GitHub events. Just as for **GitVersion** in *Chapter 6*, it can be used to automate the versioning of your containers. It also supports semantic versioning:

```
- name: Docker meta
  id: meta
  uses: docker/metadata-action@v5
  with:
    images: |
      ${{ env.REGISTRY }}/${{ env.IMAGE_NAME }}
    tags: |
      type=ref,event=branch
      type=ref,event=pr
      type=semver,pattern={{version}}
      type=semver,pattern={{major}}.{{minor}}
```

In our example, we just use the Git SHA to be able to deploy every commit, but you can easily extend the versioning according to your workflow.

Using OIDC to securely deploy to any cloud

In this recipe, we will set up our Kubernetes cluster in Azure, and we will configure OIDC in Azure to deploy to the cluster without using stored secrets.

Getting ready...

Make sure you have a PAT with at least read access to packages.

If you are experienced in Azure and you have the **Azure CLI** (`https://docs.microsoft.com/cli/azure/install-azure-cli?view=azure-cli-latest`) installed locally, then you can work from there. If you are new to Azure or you don't have the CLI installed, just use **Azure Cloud Shell** at `https://shell.azure.com`.

Set the PAT token as an environment variable:

```
$ export GHCR_PAT=<YOUR_PAT_TOKEN>
```

The token will be used by Kubernetes to read from the **GitHub Package Registry**. Open the script `setup-azure.sh` and adjust the `location` variable at the top of the file to the Azure region of your choice. You can get a list of regions using `az account list-locations -o table`. Commit and push your changes, and then run the script:

```
$ git clone https://github.com/{OWNER}/release-recipe.git
$ cd release-recipe
$ chmod +x setup-azure.sh
$ ./setup-azure.sh
```

This will create an Azure Kubernetes Service and connect it with the GitHub Container registry. While the script runs, you can configure OIDC to access it from the workflow.

How to do it...

1. Use Cloud Shell or a local terminal and create a new **app registration**:

    ```
    $ az ad app create --display-name release-recipe
    ```

2. Then create a **service principle** using app ID from the registration output:

    ```
    $ az ad sp create --id <appId>
    ```

3. Then, open the Azure portal, and in **Microsoft Entra**, find `release-recipe` under **App registrations**. Add the OIDC trust under **Certificates & secrets | Federated credentials | Add credentials**. Fill out the form. Set the organization to your GitHub username, enter the repository name, and pick **Environment** as the entity type (see *Figure 7.1*):

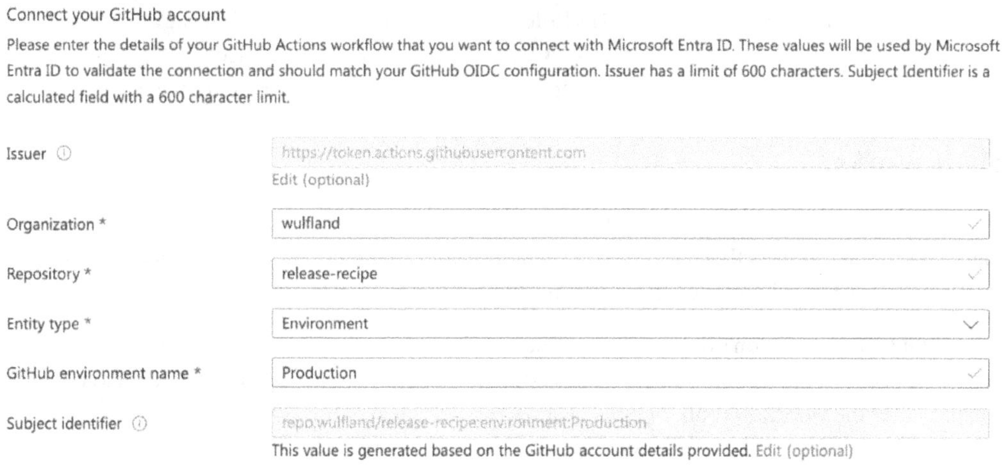

Connect your GitHub account

Please enter the details of your GitHub Actions workflow that you want to connect with Microsoft Entra ID. These values will be used by Microsoft Entra ID to validate the connection and should match your GitHub OIDC configuration. Issuer has a limit of 600 characters. Subject Identifier is a calculated field with a 600 character limit.

Issuer ⓘ	https://token.actions.githubusercontent.com
	Edit (optional)
Organization *	wulfland
Repository *	release-recipe
Entity type *	Environment
GitHub environment name *	Production
Subject identifier ⓘ	repo:wulfland/release-recipe:environment:Production
	This value is generated based on the GitHub account details provided. Edit (optional)

Figure 7.1 – Connecting your GitHub account in Microsoft Entra

Give the credentials a name and click **Add**. Note the **Application (client) ID** and **Directory (tenant) ID** of the `release-recipe` application (see *Figure 7.2*). You will need that later:

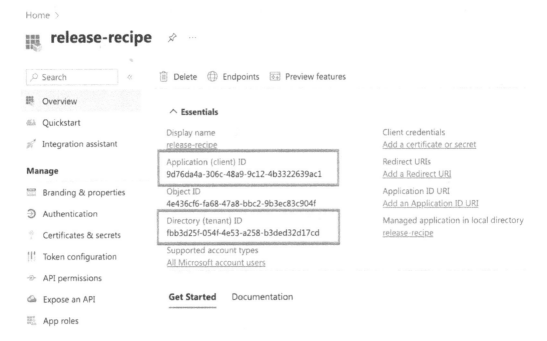

Figure 7.2 – Client and Tenant IDs from the app registration

4. Then, assign the service principle a role in your subscription. Open the subscription in the portal. Under **Access control (IAM)** | **Role assignment** | **Add** | **Add role assignment**, follow the wizard. Select role—for example, **Contributor**—and click **Next**. Select `User, group, or service principal`, and select the service principle you created earlier.

How it works...

Instead of using credentials stored as secrets to connect to a cloud provider, such as Azure, AWS, GCP, or HashiCorp, you can use OIDC. OIDC will exchange short-lived tokens for authentication instead of credentials. Your cloud provider also needs to support OIDC on their end.

When using OIDC, you don't have to store cloud credentials in GitHub, you have more granular control over what resources the workflow can access, and you have rotating, short-lived tokens that will expire after the workflows run. *Figure 7.3* shows an overview of how OIDC works:

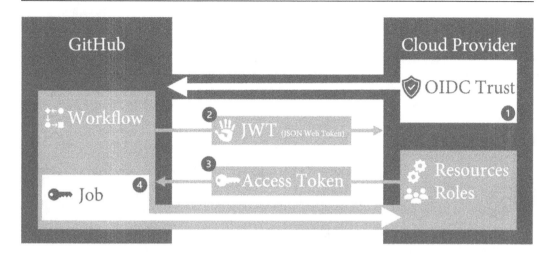

Figure 7.3 – OIDC integration with a cloud provider

The steps are the following:

1. Create an **OIDC trust** between your cloud provider and GitHub. Limit the trust to an organization and repo and further limit access to an environment, branch, or pull request.

2. The GitHub OIDC provider **auto-generates a JSON web token** during a workflow run. The token contains multiple claims to establish a secure and verifiable identity for the specific workflow job.

3. The cloud provider validates the claims and provides a **short-lived access token** that is available only for the lifetime of the job.

4. The access token is used to access resources that the identity has access to.

You can use the identity to directly access resources, or you can use it to get credentials from a secure vault (such as **Azure Key Vault** or **HashiCorp Vault**). In this way, you can safely connect to services that do not support OIDC and automated secret rotation by using the vault.

In GitHub, you can find instructions on configuring OIDC for AWS, Azure, and GDP at `https://docs.github.com/en/actions/deployment/security-hardening-your-deployments`.

Environment approval checks

We have already used environments in *Chapter 5, Automate tasks in GitHub with GitHub Actions*, so this is a repetition. Therefore, I'll keep the recipe rather short. Environments manage core releases, and we will use them in the following recipes to display the URL of our service in Kubernetes.

Getting ready...

Make sure you have the **Application (client) ID**, **Directory (tenant) ID**, and **Subscription ID** from the previous chapters at hand. The subscription ID can be obtained by using the following:

```
$ az account show
```

How to do it...

1. In the settings of your repositories, go to **Environments**, click **New environment**, and create a new environment: `Production`.
2. Add `main` as the deployment branch.
3. Add a new **Environment secret** called `AZURE_CLIENT_ID` and set it to the **Application (client) ID**.
4. Add a new **Environment secret** called `AZURE_TENANT_ID` and set it to the **Directory (tenant) ID**.
5. Add a new **Environment secret** called `AZURE_SUBSCRIPTION_ID` and set it to the **Subscription ID**.
6. Add a new **Environment secret** called `AZURE_CLUSTER_NAME` and set it to the name of the cluster (`AKSCluster` if you did not modify the `setup-azure.sh` script).
7. Add a new **Environment secret** called `AZURE_RESOURCE_GROUP` and set it to the name of the resource group (`AKSCluster` if you did not modify the `setup-azure.sh` script).

We will use these environment secrets in the next recipe to securely deploy to Kubernetes in the cloud.

How it works...

Environments add a layer of abstraction over a job in a workflow, and they can be protected by rules. See *Chapter 5*, *Automate Tasks in GitHub with GitHub Actions*, for more details on approval checks. Environments can also be trusted by OIDC entities, and this is what we are going to use in the next recipe.

Releasing the container application to AKS

Now, it is time to release our application to the production environment in AKS.

Getting ready...

Open the file `.github/workflows/publish.yml`.

How to do it...

1. Add two more environment variables to the top of the workflow:

```
env:
  REGISTRY: 'ghcr.io'
  IMAGE_NAME: '${{ github.repository }}'
  APP_NAME: 'release-recipe-app'
  SERVICE_NAME: 'release-recipe-service'
```

Add an output to the job `build-and-push-image` from the previous recipe so that the new job will be able to access the image name:

```
outputs:
  image_tag: ${{ fromJSON(steps.meta.outputs.json).tags[0] }}
```

2. Add a second job called `production` to the workflow, which only runs on pushes to `main` and is associated with the production environment. Set the URL of the environment to the output of a step we'll add later. The job will need the permissions `id-token: write` and `content: read` for OIDC to work:

```
production:
  if: github.ref == 'refs/heads/main' && github.event_name ==
'push'
  needs: build-and-push-image
  runs-on: ubuntu-latest
  permissions:
    id-token: write
    contents: read
  environment:
    name: Production
    url: ${{ steps.get-service-url.outputs.SERVICE_URL}}
```

3. Add steps to check out the repository and log into Azure using OIDC:

```
- name: Checkout
  uses: actions/checkout@v4

- name: 'Az CLI login'
  uses: azure/login@v1
  with:
    client-id: ${{ secrets.AZURE_CLIENT_ID }}
    tenant-id: ${{ secrets.AZURE_TENANT_ID }}
    subscription-id: ${{ secrets.AZURE_SUBSCRIPTION_ID }}
```

4. Set the context for the Kubernetes deployments:

```
- name: 'Az CLI set AKS context'
  uses: azure/aks-set-context@v3
  with:
    cluster-name: ${{ secrets.AZURE_CLUSTER_NAME }}
    resource-group: ${{ secrets.AZURE_RESOURCE_GROUP }}
```

5. Inspect the file `service.yml`. It deploys a LoadBalancer that displays port `3000` of our application on port `80`. Additionally, inspect `deployment.yml`, which contains the definition of our application. To replace the environment variables in the files, we use `envsubst`. We then pipe the result to `kubectl` and apply the manifest files:

```
- name: Deploy
  env:
    IMAGE: ${{ needs.build-and-push-image.outputs.image_tag }}
  run: |-
    envsubst < service.yml | kubectl apply -f -
    envsubst < deployment.yml | kubectl apply -f -
```

6. Get the URL of the service in Kubernetes using `kubectl describe service` and set it as a step output to be available in the production environment URL:

```
- name: 'Get Service URL'
  id: get-service-url
  run: |
    IP=$(kubectl describe service $SERVICE_NAME| grep
"LoadBalancer Ingress: " | awk '{print $3}')
    echo "SERVICE_URL=http://$IP" >> $GITHUB_OUTPUT
```

7. Finally, we want to check if the deployment was successful. If the application would have a `/health` endpoint, then we would query that, but because out app is very simple, we'll just rely on the returned status code:

```
- name: 'Run smoke test'
  env:
    SERVICE_URL: ${{ steps.get-service-url.outputs.SERVICE_URL}}
  run: |
    status=`curl -s --head $SERVICE_URL | head -1 | cut -f 2
-d' '`
    if [ "$status" != "200" ]
    then
      echo "Wrong HTTP Status. Actual: '$status'"
      exit 1
    fi
```

What this snippet does is query the header of the website with `curl --head`. The `-s` switch suppresses other output. It then takes the first line using `head -1`. The line looks like `HTTP/1.1 200 OK`. We cut the string by using blanks and take the second element (the status code). If the status code is not `200` (OK), it raises an exception.

8. Commit and push your changes to the main branch. This will trigger the workflow; this will push a new version of the container to the registry and publish it from there in AKS. Follow the URL of the service (see *Figure 7.4*) and verify that the website is displayed correctly:

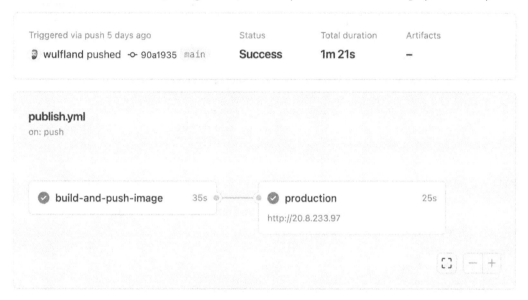

Figure 7.4 – Deploy to dynamic environments

9. It will display the **Hello World** application from your container.

How it works...

The Azure Login action will authenticate to Azure using the OIDC identity. This identity can then be granted fine-grained access to Azure resources. As we limited access to the Production environment in our repository, only this job can use the application to authenticate to Azure. You could also use branches, tags, or pull requests for that.

We then use the `aks-set-context` action to configure access to AKS using `kubectl`. This will allow us to deploy the actual application to Kubernetes.

There's more...

Kubernetes can get very complex very fast. In the real world, you are going to add DNS and SSL to the cluster and use namespaces to manage multiple containers running in parallel. This is a nice way to deploy every pull request to a dynamic environment. However, this is outside the scope of this book.

If you want to use other cloud providers instead of Azure to release the container, you'll find a hands-on for deploying to **AWS Elastic Container Service** (**ECS**) here: `https://github.com/wulfland/AccelerateDevOps/blob/main/ch9_release/Deploy_to_AWS_ECS.md`, or another on how to deploy to **Google Kubernetes Engine** (**GKE**) here: `https://github.com/wulfland/AccelerateDevOps/blob/main/ch9_release/Deploy_to_GKE.md`.

Automating the update of your dependencies

Now that we have an end-to-end workflow from our package repo into the release repo and, from there, into production, I want to show you how you can use **dependabot** together with GitHub Actions to automate the update process of your dependencies.

Getting ready...

In the repository, navigate to **Settings | Code security and analysis** and make sure that **Dependency graph** is enabled (see *Figure 7.4*):

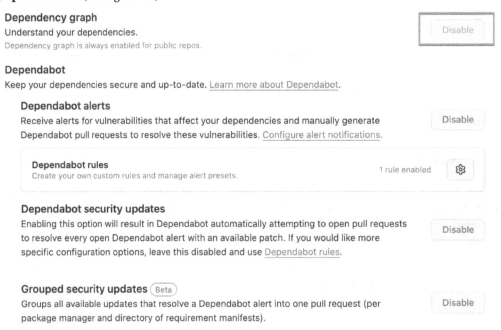

Figure 7.5 – Enabling the dependency graph and optional dependabot alerts

This will analyze your repository and detect all dependencies that you can inspect under **Insights | Dependency graph**. You can also enable **Dependabot alerts**. In this case, dependabot will notify you when there are known vulnerabilities in one of your dependencies. Dependabot security updates go one step further, and dependabot will generate a pull request with a version update to a nonvulnerable version for you. To reduce the number of pull requests, you can group the updates together (this feature is still in beta).

How to do it...

1. Create a new dependabot secret called PAT and set it with the value of the PAT token with read access to packages:

    ```
    $ gh secret set PAT --app dependabot
    ```

2. Create a new file: .github/dependabot.yml. It always starts with version:2:

    ```
    version: 2
    ```

3. Configure a new npm-registry using PAT, pointing to https://npm.pkg.github.com:

    ```
    registries:
      npm-pkg:
        type: npm-registry
        url: https://npm.pkg.github.com
        token: ${{ secrets.PAT }}
        replaces-base: true
    ```

4. The version updates are configured under updates. Add the ecosystem npm and point it to the name of the registry you created:

    ```
    updates:
      - package-ecosystem: "npm"
        directory: "/"
        schedule:
          interval: "weekly"
        registries:
          - npm-pkg
    ```

    This will check for weekly updates of npm packages, including those in the private registry.

5. Optionally, add updates for GitHub actions:

    ```
      - package-ecosystem: "github-actions"
        directory: "/"
        schedule:
          interval: "weekly"
    ```

    This will check the GitHub actions for updated versions.

6. You can also add Docker as an ecosystem:

    ```
    - package-ecosystem: "docker"
      directory: "/"
      schedule:
        interval: "weekly"
    ```

    This also works with Kubernetes manifest files, and dependabot will check for updates in image tags inside the manifest file. However, in our case, we use environment variables to directly deploy the new version.

7. Commit and push the file. Then, head over to **Insights | Dependency graph | Dependabot**. There is an entry for every ecosystem configured, and you can inspect the logfile by clicking on the link on the right side (see *Figure 7.6*):

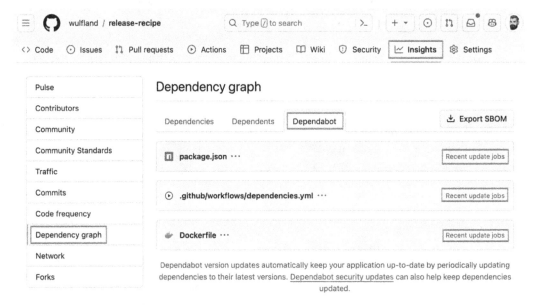

Figure 7.6 – Inspecting the logs for dependabot version updates

Check that there are no errors in the logs.

8. Now, go to the `package-recipe` repository and create a new release with a new patch version. Once the new package version is published, head over to **Insights | Dependency graph | Dependabot** and click on the link in the `package.json` line. Hit **Check for updates** to enforce dependabot to now check for updates (see *Figure 7.7*):

Dependency graph

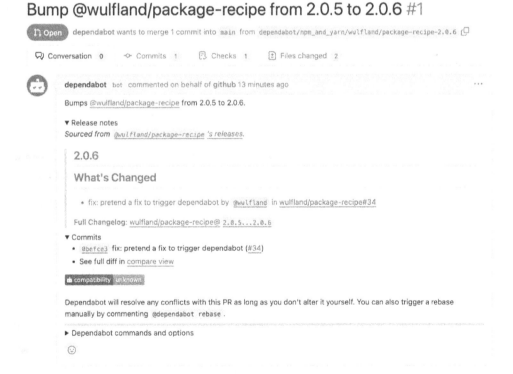

Figure 7.7 – Have dependabot now check for updates

9.   Dependabot will create a new pull request with a version update to the new version (see *Figure 7.8*):

Figure 7.8 – A pull request created by dependabot version updates

Merge the pull request or have dependabot do it by commenting @dependabot squash and merge on the pull request.

10. As we trust the owner of the package, we can automate this last step and just merge every pull request for specific versions after all checks are successful. Create a new file: .github/workflows/dependencies.yml. The workflow will run on pull_request_target, and it will need write permissions for the pull requests:

```
name: Dependabot auto-merge

on: [ pull_request_target ]

permissions:
  pull-requests: write
  contents: write
```

11. Only run the job if the author of the pull request is dependabot:

```
jobs:
  dependabot:
    runs-on: ubuntu-latest
    if: ${{ github.actor == 'dependabot[bot]' }}
    steps:
```

12. The first step is to fetch dependabot metadata:

```
- name: Dependabot metadata
  id: metadata
  uses: dependabot/fetch-metadata@v1
  with:
    github-token: "${{ secrets.GITHUB_TOKEN }}"
```

13. Next, add a step that runs on the following conditions based on the metadata: if the dependency name contains your package (replace {OWNER} with your user name) and the version update is a patch version. Merge the pull request if all checks have succeeded using gh merge --auto:

```
- name: Enable auto-merge for all patch versions
  if: ${{contains(steps.metadata.outputs.dependency-names, '@
{OWNER}/package-recipe') && steps.metadata.outputs.update-type
== 'version-update:semver-patch'}}
  run: gh pr merge --auto --merge "$PR_URL"
  env:
    PR_URL: ${{github.event.pull_request.html_url}}
    GITHUB_TOKEN: ${{secrets.GITHUB_TOKEN}}
```

Please note that this workflow will **not** trigger the `publish.yml` workflow listening for the push trigger for `main`. This is because merge is carried out using `GITHUB_TOKEN`. You can either use a PAT token, as in the last step, or you can move the publish logic to a reusable workflow and call it from this workflow directly.

14. Now, go back the `package-recipe` repository and create a new release with a new patch version. Once the new package version is published, head over to **Insights | Dependency graph | Dependabot** and click on the link in the `package.json` line. Hit **Check for updates** to enforce dependabot to now check for updates (see *Figure 7.7*). Dependabot will create a pull request, trigger the workflow, and the pull request will automatically be merged.

How it works...

Dependabot can help you to keep your dependencies up to date with less effort. You can use it to automate the update process and to keep up with the latest releases of all your dependencies.

There are many ecosystems supported by this:

- Bundler
- Cargo
- Composer
- **Dev containers** (including GitHub Codespaces)
- **Docker**
- Hex
- Elm-packages
- **Git submodules**
- **GitHub Actions**
- Go modules
- Maven and Gradle
- npm
- NuGet
- pip, pipenv, and pip-compile
- pnpm
- poetry
- pub
- Swift

- **Terraform**

- yarn

For a complete list, see the following link: `https://docs.github.com/en/code-security/dependabot/dependabot-version-updates/about-dependabot-version-updates`.

Dependabot will create pull requests for each version update of each dependency. You can use @ `Dependabot` commands on the pull request to interact with dependabot and to tell it certain things, such as ignoring versions or rebasing changes.

The `dependabot.yml` file has many options. You can specify what kind of updates are allowed, customize the commit message, group updates together, ignore certain dependencies, and add reviewers, labels, or assignees. For a complete list of configuration options, see the following link: `https://docs.github.com/en/code-security/dependabot/dependabot-version-updates/configuration-options-for-the-dependabot.yml-file`. In combination with workflows and the dependabot/fetch metadata actions, this a very powerful tool to automate your supply chain across many repositories and teams. If you look at *Chapters 6* and *7*, we use **Conventional Commits** with **GitVersion** to automate the semantic versioning based on a conventional commit message completely. We can then leverage dependabot to fully automate the update of downstream dependents.

There's more...

In our example, we directly build the container and push the deployment to Kubernetes. However, you can also use dependabot to update Kubernetes manifest files by adding an entry to the Docker `package-ecosystem` element of your `dependabot.yml` file for each directory containing a manifest, which references Docker image tags. Kubernetes manifests can be normal Kubernetes deployment files, and **Helm charts** are also supported. For more information about configuring your `dependabot.yml` file for Kubernetes, see the following link: `https://docs.github.com/en/code-security/dependabot/dependabot-version-updates/configuration-options-for-the-dependabot.yml-file#docker`.

Clean up

Please don't forget to delete your cluster when you are done with the recipes so that no unnecessary costs will occur. You can use the script `destroy-azure.sh` in the repo and run it locally or just run the following command in the **Azure Cloud Shell** (`https://shell.azure.com`):

```
$ az group delete --resource-group AKSCluster --yes
```

Just check that you did not change the name of the resource group when setting everything up.

Summary

Congratulations on making it to the end of the book. I hope the practical, hands-on, focused recipes helped you to build a foundation for automating all kinds of tasks in your day-to-day work and helped increase the productivity in your engineering teams. I tried to include all the aspects of GitHub Actions that are relevant, balancing simplicity and real-world applicability. GitHub is a platform that is evolving very fast, with changes being released multiple times a day. If you consider the partners and actions from the open source community, the GitHub ecosystem is huge and changes all the time. If you have encountered changes during the hands-on labs in this book, please reach out to me on GitHub by creating an issue or submitting a pull request, and I will try to incorporate the changes in the repository. I hope you enjoyed the book, and I hope you will enjoy GitHub Actions as I do; it is the best automation platform I've ever used.

Index

Y

YAML Ain't Markup Language

Other Books You May Enjoy

If you enjoyed this book, you may be interested in these other books by Packt:

Mastering GitHub Actions

Eric Chapman

ISBN: 978-1-80512-862-5

- Explore GitHub Actions' features for team and business settings
- Create reusable workflows, templates, and standardized processes to reduce overhead
- Get to grips with CI/CD integrations, code quality tools, and communication
- Understand self-hosted runners for greater control of resources and settings
- Discover tools to optimize GitHub Actions and manage resources efficiently
- Work through examples to enhance projects, teamwork, and productivity

Automating DevOps with GitLab CI/CD Pipelines

Christopher Cowell, Nicholas Lotz, Chris Timberlake

ISBN: 9781803233000

- Gain insights into the essentials of Git, GitLab, and DevOps

- Understand how to create, view, and run GitLab CI/CD pipelines

- Explore how to verify, secure, and deploy code with GitLab CI/CD pipelines

- Configure and use GitLab Runners to execute CI/CD pipelines

- Explore advanced GitLab CI/CD pipeline features like DAGs and conditional logic

- Follow best practices and troubleshooting methods of GitLab CI/CD pipelines

- Implement end-to-end software development lifecycle workflows using examples

Packt is searching for authors like you

If you're interested in becoming an author for Packt, please visit `authors.packtpub.com` and apply today. We have worked with thousands of developers and tech professionals, just like you, to help them share their insight with the global tech community. You can make a general application, apply for a specific hot topic that we are recruiting an author for, or submit your own idea.

Share your thoughts

Now you've finished *GitHub Actions Cookbook*, we'd love to hear your thoughts! Scan the QR code below to go straight to the Amazon review page for this book and share your feedback or leave a review on the site that you purchased it from.

https://packt.link/r/1835468942

Your review is important to us and the tech community and will help us make sure we're delivering excellent quality content.

Download a free PDF copy of this book

Thanks for purchasing this book!

Do you like to read on the go but are unable to carry your print books everywhere?

Is your eBook purchase not compatible with the device of your choice?

Don't worry, now with every Packt book you get a DRM-free PDF version of that book at no cost.

Read anywhere, any place, on any device. Search, copy, and paste code from your favorite technical books directly into your application.

The perks don't stop there, you can get exclusive access to discounts, newsletters, and great free content in your inbox daily

Follow these simple steps to get the benefits:

1. Scan the QR code or visit the link below

https://packt.link/free-ebook/9781835468944

2. Submit your proof of purchase
3. That's it! We'll send your free PDF and other benefits to your email directly